T0269545

Derivative with a New Parameter
Theory, Methods and Applications

Derivative with a New Parameter
Theory, Methods and Applications

Abdon Atangana
Institute for Groundwater Studies,
University of the Free State,
Bloemfontein, South Africa

AMSTERDAM • BOSTON • HEIDELBERG • LONDON
NEW YORK • OXFORD • PARIS • SAN DIEGO
SAN FRANCISCO • SINGAPORE • SYDNEY • TOKYO
Academic Press is an imprint of Elsevier

Academic Press is an imprint of Elsevier
125 London Wall, London, EC2Y 5AS, UK
525 B Street, Suite 1800, San Diego, CA 92101-4495, USA
225 Wyman Street, Waltham, MA 02451, USA
The Boulevard, Langford Lane, Kidlington, Oxford OX5 1GB, UK

© 2016 Elsevier Ltd. All rights reserved.

No part of this publication may be reproduced or transmitted in any form or by any means, electronic or mechanical, including photocopying, recording, or any information storage and retrieval system, without permission in writing from the publisher. Details on how to seek permission, further information about the Publisher's permissions policies and our arrangements with organizations such as the Copyright Clearance Center and the Copyright Licensing Agency, can be found at our website: www.elsevier.com/permissions.

This book and the individual contributions contained in it are protected under copyright by the Publisher (other than as may be noted herein).

Notices
Knowledge and best practice in this field are constantly changing. As new research and experience broaden our understanding, changes in research methods, professional practices, or medical treatment may become necessary.

Practitioners and researchers must always rely on their own experience and knowledge in evaluating and using any information, methods, compounds, or experiments described herein. In using such information or methods they should be mindful of their own safety and the safety of others, including parties for whom they have a professional responsibility.

To the fullest extent of the law, neither the Publisher nor the authors, contributors, or editors, assume any liability for any injury and/or damage to persons or property as a matter of products liability, negligence or otherwise, or from any use or operation of any methods, products, instructions, or ideas contained in the material herein.

ISBN: 978-0-08-100644-3

Library of Congress Cataloging-in-Publication Data
A catalog record for this book is available from the Library of Congress

British Library Cataloguing in Publication Data
A catalogue record for this book is available from the British Library

For information on all Academic Press publications
visit our website at http://store.elsevier.com/

ELSEVIER   Book Aid
            International

Working together
to grow libraries in
developing countries

www.elsevier.com • www.bookaid.org

I dedicate this book to the Lord Jesus Christ, who died on the cross for our sins. Also to the memory of my late mother Ngono Antoinette, who died in 2010.

CONTENTS

And the lord God said, "Let differential calculus exist," and the differential calculus existed within the realm of God's knowledge. And the lord God said, "Who should I send to introduce this concept to humankind?" and Isaac Barrow, Rene Descartes, Christian Huygens, Blaise Pascal, John Wallis, Isaac Newton, and Gottfried Leibniz said, "lord God please send us."

The primary objects of study in differential calculus are the derivative of a function, related notions such as the differential, and their applications. The derivative of a function at a chosen input value describes the rate of change of the function near that input value. Differentiation has applications to nearly all quantitative disciplines. For example, in physics, the derivative of the displacement of a moving body with respect to time is the velocity of the body, and the derivative of velocity with respect to time is acceleration. The derivative of the momentum of a body equals the force applied to the body; rearranging this derivative statement leads to the famous $F = ma$ equation associated with Newton's second law of motion. The reaction rate of a chemical reaction is a derivative. In operations research, derivatives determine the most efficient ways to transport materials and design factories. The notion is the concept mainly used in modeling real-world problems. Several definitions have been proposed in the literature, ranging from local derivative to variable order derivative. All of them have advantages and limitations. With the local derivative, many real-world problems cannot be replicated accurately. For derivatives with fractional order, some basic properties of derivative cannot be satisfied, therefore some important classes of real-world problems cannot be modeled—for instance, the boundaries layers problem. In this book, we present a new derivative, which is defined as a local derivative but has a fractional order. We present in Chapter 1 the history of derivative from the local derivative to fractional derivative, together with their advantages and limitations. Lists of criteria that need to be satisfied by an operator to be considered as a fractional derivative are presented. In Chapter 2, we present the definition of the new derivative and the motivation of this particular definition. A theory associated to the new derivative is presented. In Chapter 3, we present three novel integral transform methods and their properties. In Chapter 4, we present analytical and numerical methods to solve ordinary and partial differential equations with the new derivative. Finally, in Chapter 5, we present the possible application of the new derivative to groundwater flow and epidemiology problems.

ACKNOWLEDGMENTS

I would like to thank the Lord God Jesus Christ for giving me life, strength and courage to complete this book.

Then the lord God said, "It is not good that Abdon Atangana should be alone; I will make him a helper fit for him." And the rib that the lord God had taken from Abdon Atangana, he made into a beautiful, caring, loving, wise, intelligent woman (Ernestine Alabaraoye) and brought her to Abdon Atangana. Then Abdon Atangana said, "This at last is bone of my bones and flesh of my flesh; she shall be called madam Atangana, because she was taken out of me. Therefore Abdon Atangana shall leave his father and his mother and hold fast to his God-fearing wife Ernestine Alabaraoye, and they shall become one flesh." I would like to thank my lovely wife Ernestine Alabaraoye for her love, good food, prayers, and constant support.

To my father: Dr. Pierre Ndzengue, the ambassador of Cameroon to Japan, South Korea, Australia and New Zealand for his prayers and financial support. Tara Noah Jean for his constant prayers and faith.

To my aunt and uncle, Prof. Mbanya Dora and Claude, for their encouragement and their prayers.

To my family: Eleme Catherine, Ngono Adelaide, Ngono Antoinette, Eyebe Isidore, maitre Ebeme Ateba Joseph, and my grandmother Ngono Adelaide for their constant support and prayers.

To my friends Dr. Emile Franck Goufo, Prof. Dr. Xiao-Jun Yang, Prof Dr. Dumitru Baleanu, and Prof. Dr. Suares Clovis Noutchi Oukouomie for their academic support.

To my PhD supervisor Prof. Dr. J.F. Botha for his advice and academic support.

To the Dean of the Faculty of Natural and Agricultural Sciences Prof. Dr. Niel Heideman for his kind advice and academic support.

To Pastor David Kennedy for his spiritual advices and prayers.

A special thank to the Claude Leon Foundation, which supported this project financially.

CHAPTER 1

History of derivatives from Newton to Caputo

1.1 INTRODUCTION

In mathematics, differential calculus is a subfield of calculus concerned with the study of the rates at which quantities change. It is one of the two traditional divisions of calculus, the other being integral calculus. The primary objects of study in differential calculus are the derivative of a function, related notions such as the differential, and their applications. The derivative of a function at a chosen input value describes the rate of change of the function near that input value [1]. The process of finding a derivative is called differentiation. Geometrically, the derivative at a point is the slope of the tangent line to the graph of the function at that point, provided that the derivative exists and is defined at that point. For a real-valued function of a single real variable, the derivative of a function at a point generally determines the best linear approximation to the function at that point. The concept of a derivative in the sense of a tangent line is a very old one, familiar to Greek geometers such as Euclid (*c.* 300 BC), Archimedes (*c.* 287–212BC) [2], and Apollonius of Perga (*c.* 262–190BC) [3–5]. Archimedes also introduced the use of infinitesimals, although these were primarily used to study areas and volumes rather than derivatives and tangents; see Archimedes' use of infinitesimals.

The use of infinitesimals to study rates of change can be found in Indian mathematics, perhaps as early as AD500, when the astronomer and mathematician Aryabhata (476–550) used infinitesimals to study the motion of the moon. The use of infinitesimals to compute rates of change was developed significantly by Bhskara II (1114–1185); indeed, it has been argued that many of the key notions of differential calculus can be found in his work, such as "Rolle's theorem." The Persian mathematician, Sharaf al-Dn al-Ts (1135–1213) [5], was the first to discover the derivative of cubic polynomials, an important result in differential calculus; his Treatise on Equations developed concepts related to differential calculus, such as the derivative function and the maxima and minima of curves, in order to solve cubic equations which may not have positive solutions. The modern development of calculus is usually credited to Isaac Newton (1643–1727) [6],

Derivative with a New Parameter. http://dx.doi.org/10.1016/B978-0-08-100644-3.00001-5
© 2016 Elsevier Ltd. All rights reserved.

and Gottfried Leibniz (1646–1716) [7], who provided independent and unified approaches to differentiation and derivatives. The key insight, however, that earned them this credit, was the fundamental theorem of calculus relating differentiation and integration: this rendered obsolete most previous methods for computing areas and volumes, which had not been significantly extended since the time of Ibn al-Haytham (Alhazen). For their ideas on derivatives, both Newton and Leibniz built on significant earlier work by mathematicians such as Isaac Barrow (1630–1677), Ren Descartes (1596–1650), Christian Huygens (1629–1695), Blaise Pascal (1623–1662), and John Wallis (1616–1703) [2, 8–12]. Isaac Barrow is generally given credit for the early development of the derivative. Nevertheless, Newton and Leibniz remain key figures in the history of differentiation, not least because Newton was the first to apply differentiation to theoretical physics, while Leibniz systematically developed much of the notation still used today.

Since the seventeenth century, many mathematicians have contributed to the theory of differentiation. In the nineteenth century, calculus was put on a much more rigorous footing by mathematicians such as Augustin Louis Cauchy (1789–1857), Bernhard Riemann (1826–1866), and Karl Weierstrass (1815–1897) [13–15]. It was also during this period that the differentiation was generalized to Euclidean space and the complex plane. This concept used in almost all the branches of sciences for modeling. However, due to the complexities associated with the physical problems encountered in nature, the concept of derivative was modified. Fractional calculus, in the understanding of its theoretical and real-world presentations in numerous regulations (e.g., astronomy and manufacturing problems), is discovered to be accomplishing of pronouncing phenomena owning long-range memory special effects that are challenging to handle through traditional integer-order calculus. Nearby an increasing concentration has been in the modification of fractional calculus as a successful modeling instrument for complicated systems, contributing to innovative viewpoints in their dynamical investigation and regulator. This improvement in the methodical knowledge is established by an enormous quantity of events developing on the subject, manuscripts, and presentations in the past years. Nevertheless, countless singularities still pose significant challenges to the apprehensive population and fractional calculus appears to be a very useful mathematical tool, particularly, the advantages of the concept of fractional calculus and their associated models via differential partial differential equations linking fractional order integro-differential operators and their appliances have been previously intensively investigated with great success during the last years. We shall present in the next section different definitions of derivative that are found in the literature.

1.2 DEFINITION OF LOCAL AND FRACTIONAL DERIVATIVE

Since the introduction of the formulation of derivative by Newton, several others' definitions have been introduced. We shall in this section present these definitions from Newton to Caputo

Definition 1.2.1. Let f be a function defined in a closed interval $[a, b]$, then the derivative of a function $f(x)$ is written as $f'(x)$:

$$f'(x) = \lim_{h \to 0} \frac{f(x+h) - f(x)}{h}. \tag{1.1}$$

There are a few different notations used to refer to derivatives. It is very important that one learns to identify these different ways of denoting the derivative and that one is consistent in usage of them. The notation for derivatives introduced by Gottfried Leibniz is one of the earliest. It is still commonly used when the equation $y = f(x)$ is viewed as a functional relationship between dependent and independent variables. Then the first derivative is denoted by

$$\frac{dy}{dx}, \quad \frac{df}{dx}(x), \quad \text{or} \quad \frac{d}{dx}f(x), \tag{1.2}$$

and was once thought of as an infinitesimal quotient. Higher derivatives are expressed using the notation:

$$\frac{d^n y}{dx^n}, \quad \frac{d^n f}{dx^n}(x), \quad \text{or} \quad \frac{d^n}{dx^n}f(x), \tag{1.3}$$

for the nth derivative of $y = f(x)$ (with respect to x). Sometimes referred to as prime notation, one of the most common modern notations for differentiation is due to Joseph–Louis Lagrange and uses the prime mark, so that the derivative of a function $f(x)$ is denoted $f'(x)$ or simply f'. Similarly, the second and third derivatives are denoted:

$$(f')' = f'' \quad \text{and} \quad (f'')' = f'''. \tag{1.4}$$

To denote the number of derivatives beyond this point, some authors use Roman numerals in superscript, whereas others place the number in parentheses:

$$f^{\text{iv}} \quad \text{or} \quad f^4. \tag{1.5}$$

The latter notation generalizes to yield the notation $f^{(n)}$ for the nth derivative of f. This notation is most useful when we wish to talk about the derivative as being a function itself, as in this case, the Leibniz notation can become cumbersome. Newton's notation for differentiation, also called the dot notation, places a dot over the function name to represent a time derivative. If $y = f(t)$, then

$$\dot{y} \quad \text{and} \quad \ddot{y}, \tag{1.6}$$

denoting, respectively, the first and second derivatives of y with respect to t. This notation is used exclusively for time derivatives, meaning that the independent variable of the function represents time. Euler's notation uses a differential operator D, which is applied to a function f to give the first derivative Df. The second derivative is denoted D^2f, and the nth derivative is denoted $D^n f$. If $y = f(x)$ is a dependent variable, then often the subscript x is attached to the D to clarify the independent variable x. Euler's notation is then written:

$$D_x y \quad \text{or} \quad D_x f(x), \tag{1.7}$$

although this subscript is often omitted when the variable x is understood, for instance, when this is the only variable present in the expression. Euler's notation is useful for stating and solving linear differential equations. Beside the concept of local derivative, the notion of fractional order derivative has attracted attention of many scholars around the world. There exists a vast literature on different definitions of fractional derivatives. The most popular ones are the Riemann–Liouville and the Caputo derivatives.

Definition 1.2.2. Unlike classical Newtonian derivatives, a fractional derivative is defined via a fractional integral. Let f be a function defined in an opened interval $(0, a)$ then, the fractional order derivative of f is given as:

$$\frac{1}{\Gamma(\alpha)} \int_0^x (x - t)^{\alpha - 1} f(t) dt$$

The Riemann–Liouville derivative is the most-used generalization of the derivative for theoretical purpose. It is based on Cauchy's formula for calculating iterated integrals. In the case of Riemann–Liouville, we have the following definition [16–18]:

$$D^\alpha f(x) = \frac{1}{\Gamma(n-\alpha)} \frac{d^n}{dx^n} \int_0^x \frac{f(t)}{(x-t)^{\alpha+1-n}}\, dt. \qquad (1.8)$$

There is another option for computing fractional derivatives: the Caputo fractional derivative. M. Caputo introduced it in his 1967 paper. In contrast to the Riemann–Liouville fractional derivative, when solving differential equations using Caputo's definition, it is not necessary to define the fractional order initial conditions. Caputo's definition is illustrated as follows [19–21]:

$$_a^C D_t^\alpha f(t) = \frac{1}{\Gamma(n-\alpha)} \int_a^t \frac{f^{(n)}(\tau)\, d\tau}{(t-\tau)^{\alpha+1-n}}. \qquad (1.9)$$

Weyl introduced the following definition [19, 21]:

$$D^\alpha f(x) = \frac{1}{\Gamma(n-\alpha)} \frac{d^n}{dx^n} \int_x^\infty \frac{f(t)}{(x-t)^{\alpha+1-n}}\, dt. \qquad (1.10)$$

With the Erdelyi–Kober type we have the following definition [22]:

$$D_{0,\sigma,\eta}^\alpha (f(x)) = x^{-n\sigma} \left(\frac{1}{\sigma x^{\sigma-1}} \frac{d}{dx}\right)^n x^{\sigma(n+\eta)} I_{0,\sigma,\eta+\sigma}(f(x)), \qquad (1.11)$$

here

$$I_{0,\sigma,\eta+\sigma}(f(x)) = \frac{\sigma x^{-\sigma(\eta+\alpha)}}{\Gamma\alpha} \int_0^x \frac{t^{\sigma\eta+\sigma-1} f(t)}{(t^\sigma - x^\sigma)^{1-\alpha}}\, dt. \qquad (1.12)$$

With the Hadamard type, we have the following definition [23]:

$$D_0^\alpha (f(x)) = \frac{1}{\Gamma(n-\alpha)} \left(x\frac{d}{dx}\right)^n \int_0^x \left(\log\left(\frac{x}{t}\right)\right)^{n-\alpha-1} f(t)\frac{dt}{t}. \qquad (1.13)$$

With the Riesz type, we have the following definition [24]:

$$D_x^\alpha (f(x)) = -\frac{1}{2\cos\left(\frac{\alpha\pi}{2}\right)} \frac{1}{\Gamma(\alpha)} \left(\frac{d}{dx}\right)^m$$
$$\times \left(\int_{-\infty}^x (x-t)^{m-\alpha-1} f(t)\, dt + \int_x^\infty (x-t)^{m-\alpha-1} f(t)\, dt\right). \qquad (1.14)$$

In 1998, Davison and Essex published a paper that provides a variation to the Riemann–Liouville definition suitable for conventional initial value problems within the realm of fractional calculus. The definition is as follows [25]:

$$D_x^\alpha(f(x)) = \frac{\mathrm{d}^{n+1-k}}{\mathrm{d}x^{n+1-k}} \int_0^x \frac{(x-t)^{-\alpha}}{\Gamma(1-\alpha(x))} \frac{\mathrm{d}^k f(t)}{\mathrm{d}t^k} \, \mathrm{d}t. \tag{1.15}$$

In an article published by Coimbra in 2003, a variable-order (VO) differential operator is defined as follows [26]:

$$D_x^{\alpha(x)}(f(x)) = \frac{1}{\Gamma(1-\alpha(x))} \int_0^x (x-t)^{-\alpha(t)} \frac{\mathrm{d}f(t)}{\mathrm{d}t} + \frac{(f(0^+) - f(0^-))x^{-\alpha(t)}}{\Gamma(1-\alpha(x))}. \tag{1.16}$$

Guy Jumarie proposed a simple alternative definition to the Riemann–Liouville derivative [27]:

$$D^\alpha f(x) = \frac{1}{\Gamma(n-\alpha)} \frac{\mathrm{d}^n}{\mathrm{d}x^n} \int_0^x \frac{f(t) - f(0)}{(x-t)^{\alpha+1-n}} \, \mathrm{d}t. \tag{1.17}$$

In the area of combinatorics, the q-derivative, or Jackson derivative, is a q-analog of the ordinary derivative, introduced by Frank Hilton Jackson. It is the inverse of Jackson's q-integration [28].

Definition 1.2.3. Thus q-derivative of a function $f(x)$ is defined as:

$$\left(\frac{\mathrm{d}}{\mathrm{d}x}\right)_q f(x) = \frac{f(qx) - f(x)}{qx - x}. \tag{1.18}$$

It is also often written as $D_q f(x)$. In applied mathematics and mathematical analysis, the fractal derivative is a nonstandard type of derivative in which the variable such as t has been scaled according to t^α. The derivative is defined in fractal geometry.

Definition 1.2.4. The concept of the fractal derivative of a function $u(t)$ with respect to a fractal measure t has been introduced as follows [29]:

$$\frac{\partial f(t)}{\partial t^\alpha} = \lim_{t_1 \to t} \frac{f(t_1) - f(t)}{t_1^\alpha - t^\alpha}, \quad \alpha > 0. \tag{1.19}$$

A more general definition is given by

$$\frac{\partial^\beta f(t)}{\partial t^\alpha} = \lim_{t_1 \to t} \frac{f^\beta(t_1) - f^\beta(t)}{t_1^\alpha - t^\alpha}, \quad \alpha > 0, \beta > 0. \tag{1.20}$$

Definition 1.2.5. For a function $f: [0, 1] \to \mathbb{R}$, the limit:

$$D_+^\alpha (f(x)) = \lim_{y \to x^+} \frac{d^\alpha (f(y) - f(x))}{d(+(y-x))^\alpha}, \quad 0 < \alpha < 1 \qquad (1.21)$$

exists and is finite, then f is said to have right (Left) LFD (Local fractional derivative) of order α at $y = x$.

Definition 1.2.6. Left and right Riemann–Liouville derivative of variable fractional order: let $0 < \alpha(t, x) < 1$ for all $(x, t) \in [a, b]$, then the left Riemann–Liouville derivative of variable fractional order $\alpha(.,.)$ is given as [30]:

$$_aD_t^{\alpha(.,.)} (f(t)) = \frac{d}{dt} \int_a^t \frac{1}{\Gamma(1 - \alpha(x, t))} (t-x)^{-\alpha(x,t)} f(x) \, dx, \quad (t > a). \quad (1.22)$$

Likewise, we have the following expression referred to as the right Riemann–Liouville derivative of variable fractional order $\alpha(.,.)$:

$$_bD_t^{\alpha(.,.)} (f(t)) = \frac{d}{dt} \int_t^a \frac{1}{\Gamma(1 - \alpha(x, t))} (x-t)^{-\alpha(.,.)} f(x) \, dx, \quad (t < b). \quad (1.23)$$

Definition 1.2.7. Left and right Caputo derivative of variable fractional order: let $0 < \alpha(t, x) < 1$ for all $(x, t) \in [a, b]$, then the left Caputo derivative of variable fractional order $\alpha(.,.)$ is given as [31]:

$$_aD_t^{\alpha(.,.)} (f(t)) = \int_a^t \frac{1}{\Gamma(1 - \alpha(x, t))} (t-x)^{-\alpha(x,t)} \frac{df(x)}{dx} \, dx, \quad (t > a). \quad (1.24)$$

Likewise, we have the following expression referred to as the right Caputo derivative of variable fractional order $\alpha(.,.)$:

$$_bD_t^{\alpha(.,.)} (f(t)) = \int_t^b \frac{1}{\Gamma(1 - \alpha(x, t))} (x-t)^{-\alpha(.,.)} \frac{df(x)}{dx} \, dx, \quad (t < b). \quad (1.25)$$

Due to the complexities of the above definition of VO derivative, Abdon Atangana has proposed an alternative formulation of VO derivative based on the Riemann–Liouville and Caputo fractional derivatives. These versions are given in the next definition [31].

Definition 1.2.8. Left and right Caputo derivative of variable fractional order: Let $f: \mathbb{R} \to \mathbb{R}, x \to f(x)$ denote a continuous and necessary differentiable function, let $\alpha(x)$ be continuous function in $(0.1]$. Then its left Caputo variable fractional derivative is given as:

$$_{a}D_{t}^{\alpha(.,.)}(f(t)) = \frac{1}{\Gamma(1-\alpha(t))} \int_{a}^{t} (t-x)^{-\alpha(x)} \frac{df(x)}{dx} \, dx, \quad (t > a). \quad (1.26)$$

Likewise, we have the following expression referred to as the right Caputo derivative of variable fractional order $\alpha(.)$:

$$_{b}D_{t}^{\alpha(.,.)}(f(t)) = \frac{1}{\Gamma(1-\alpha(t))} \int_{t}^{b} (x-t)^{-\alpha(.,.)} \frac{df(x)}{dx} \, dx, \quad (t < b). \quad (1.27)$$

Definition 1.2.9. Left and right Riemann–Liouville derivative of variable fractional order: Let $f\colon \mathbb{R} \to \mathbb{R}$, $x \to f(x)$ denote a continuous and necessary differentiable function, let $\alpha(x)$ be continuous function in $(0.1]$. Then its left Riemann–Liouville variable fractional derivative is given as:

$$_{a}D_{t}^{\alpha(.,.)}(f(t)) = \frac{1}{\Gamma(1-\alpha(t))} \frac{d}{d} \int_{a}^{t} (t-x)^{-\alpha(x)} \, dx, \quad (t > a). \quad (1.28)$$

Likewise, we have the following expression referred to as the right Riemann–Liouville derivative of variable fractional order $\alpha(.)$:

$$_{b}D_{t}^{\alpha(.,.)}(f(t)) = \frac{1}{\Gamma(1-\alpha(t))} \frac{d}{dt} \int_{t}^{b} (x-t)^{-\alpha(.,.)} \, dx, \quad (t < b). \quad (1.29)$$

1.3 DEFINITIONS AND PROPERTIES OF THEIR ANTI-DERIVATIVES

In calculus, an anti-derivative, primitive integral, or indefinite integral of a function f is a differentiable function F whose derivative is equal to f, meaning $F = f$. The process of solving for anti-derivatives is called anti-differentiation (or indefinite integration) and its opposite operation is called differentiation, which is the process of finding a derivative. Anti-derivatives are related to definite integrals through the fundamental theorem of calculus: the definite integral of a function over an interval is equal to the difference between the values of an anti-derivative evaluated at the endpoints of the interval. We shall present some anti-derivatives associated to the derivatives presented earlier [5, 6, 17–19, 21].

1.3.1 Anti-derivative with integer order

Definition 1.3.1. If F is an anti-derivative of the integrable function f and f is continuous over the interval $[a, b]$, then

$$\int_a^b f(x) \, dx = F(b) - F(a). \tag{1.59}$$

Because of this, each of the infinitely many anti-derivatives of a given function f is sometimes called the "general integral" or "indefinite integral" of f and is written using the integral symbol with no bounds:

$$\int f(x) \, dx. \tag{1.60}$$

If F is an anti-derivative of f, and the function f is defined on some interval, then every other anti-derivative G of f differs from F by a constant: there exists a number C such that $G(x) = F(x) + C$ for all x. C is called the arbitrary constant of integration. The fundamental theorem of calculus states that differentiation and integration are inverse operations. More precisely, it relates the values of anti-derivatives to definite integrals. Because it is usually easier to compute an anti-derivative than to apply the definition of a definite integral, the fundamental theorem of calculus provides a practical way of computing definite integrals. It can also be interpreted as a precise statement of the fact that differentiation is the inverse of integration.

Theorem 1.3.1. *The fundamental theorem of calculus states: If a function f is continuous on the interval $[a, b]$ and if F is a function whose derivative is f on the interval (a, b), then [5]:*

$$\int_a^b f(x) \, dx = F(b) - F(a). \tag{1.61}$$

Furthermore, for every x in the interval (a, b):

$$\frac{d}{dx} \int_a^x f(t) \, dt = f(x). \tag{1.62}$$

The theorem above has two parts. Loosely put, the first part deals with the derivative of an anti-derivative, while the second part deals with the relationship between anti-derivatives and definite integrals. We shall next present the proof for parts 1 and 2. The part is sometimes referred to as the first fundamental theorem of calculus. For the first part we have the following proof [5].

Proof. Let f be a continuous real-valued function defined on a closed interval $[a, b]$. Let F be the function defined, for all x in $[a, b]$, by:

$$F(x) = \int_a^x f(t) \, dt. \tag{1.63}$$

Then, F is continuous on $[a, b]$, differentiable on the open interval (a, b), and

$$F'(x) = f(x) \tag{1.64}$$

for all $x \in (a, b)$.

Alternatively, if f is merely Riemann integrable, then F is continuous on $[a, b]$ (but not necessarily differentiable).

Corollary 1.3.1. *The fundamental theorem is often employed to compute the definite integral of a function f for which an anti-derivative F is known. Specifically, if f is a real-valued continuous function on $[a, b]$, and F is an anti-derivative of f in $[a, b]$, then*

$$\int_a^b f(t) \, dt = F(b) - F(a). \tag{1.65}$$

The corollary assumes continuity on the whole interval. This result is strengthened slightly in the following part of the theorem. □

Second part. This part is sometimes referred to as the second fundamental theorem of calculus or the Newton–Leibniz axiom.

Proof. Let f and F be real-valued functions defined on a closed interval $[a, b]$ such that the derivative of F is f. That is, f and F are functions such that for all x in $[a, b]$:

$$F'(x) = f(x). \tag{1.66}$$

If f is Riemann integrable on $[a, b]$ then

$$\int_a^b f(x) \, dx = F(b) - F(a). \tag{1.67}$$

The second part is somewhat stronger than the corollary because it does not assume that f is continuous.

When an anti-derivative F exists, then there are infinitely many anti-derivatives for f, obtained by adding to F an arbitrary constant. Also, by the first part of the theorem, anti-derivatives of f always exist when f is continuous.

For a given $f(t)$, define the function $F(x)$ as:

$$F(x) = \int_a^x f(t) \, dt. \tag{1.68}$$

For any two numbers x_1 and $x_1 + x$ in $[a, b]$, we have

$$F(x_1) = \int_a^{x_1} f(t) \, dt \tag{1.69}$$

and

$$F(x_1 + \Delta x) = \int_a^{x_1 + \Delta x} f(t) \, dt. \tag{1.70}$$

Subtracting the two equalities gives

$$F(x_1 + \Delta x) - F(x_1) = \int_a^{x_1 + \Delta x} f(t) \, dt - \int_a^{x_1} f(t) \, dt. \tag{1.71}$$

It can be shown that

$$\int_a^{x_1} f(t) \, dt + \int_{x_1}^{x_1 + \Delta x} f(t) \, dt = \int_a^{x_1 + \Delta x} f(t) \, dt. \tag{1.72}$$

The sum of the areas of two adjacent regions is equal to the area of both regions combined. Manipulating this equation gives

$$\int_a^{x_1 + \Delta x} f(t) \, dt - \int_a^{x_1} f(t) \, dt = \int_{x_1}^{x_1 + \Delta x} f(t) \, dt. \tag{1.73}$$

Substituting the above into Equation (1.71) results in

$$F(x_1 + \Delta x) - F(x_1) = \int_{x_1}^{x_1 + \Delta x} f(t) \, dt. \tag{1.74}$$

According to the mean value theorem for integration, there exists a real number $c(\Delta x)$ in $[x_1, x_1 + \Delta x]$ such that

$$\int_{x_1}^{x_1 + \Delta x} f(t) \, dt = f(c(\Delta x)) \, \Delta x. \tag{1.75}$$

To keep the notation simple, we shall continue writing c instead of $c(\Delta x)$ but one should keep in mind that c does depend on Δx. Substituting the above into Equation (1.74) we get

$$F(x_1 + \Delta x) - F(x_1) = f(c) \Delta x. \tag{1.76}$$

Dividing both sides by x gives

$$\frac{F(x_1 + \Delta x) - F(x_1)}{\Delta x} = f(c). \tag{1.77}$$

The expression on the left side of the equation is Newton's difference quotient for F at x_1. Take the limit as $\Delta x \longrightarrow 0$ on both sides of the equation:

$$\lim_{\Delta x \to 0} \frac{F(x_1 + \Delta x) - F(x_1)}{\Delta x} = \lim_{\Delta x \to 0} f(c). \tag{1.78}$$

The expression on the left side of the equation is the definition of the derivative of F at x_1:

$$F'(x_1) = \lim_{\Delta x \to 0} f(c). \tag{1.79}$$

To find the other limit, we use the squeeze theorem. The number c is in the interval $[x_1, x_1 + \Delta x]$, so $x_1 \le c \le x_1 + \Delta x$:

Also,

$$\lim_{\Delta x \to 0} x_1 = x_1 \quad \text{and} \quad \lim_{\Delta x \to 0} x_1 + \Delta x = x_1. \tag{1.80}$$

Therefore, according to the squeeze theorem:

$$\lim_{\Delta x \to 0} c = x_1. \tag{1.81}$$

Substituting into Equation (1.79), we get

$$F'(x_1) = \lim_{c \to x_1} f(c). \tag{1.82}$$

The function f is continuous at c, so the limit can be taken inside the function. Therefore, we get:

$$F'(x_1) = f(x_1), \tag{1.83}$$

which completes the proof.   □

Proof of the corollary

Proof. Suppose F is an anti-derivative of f, with f continuous on $[a, b]$. Let

$$G(x) = \int_a^x f(t) \, dt. \tag{1.84}$$

By the first part of the theorem, we know G is also an anti-derivative of f. Since $F' - G' = 0$ the mean value theorem implies that $F - G$ is a constant function, that is, there is a number c such that $G(x) = F(x) + c$, for all x in $[a, b]$. Letting $x = a$, we have

$$F(a) + c = G(a) = \int_a^a f(t) \, dt = 0, \tag{1.85}$$

which means $c = F(a)$. In other words $G(x) = F(x)F(a)$, and so

$$\int_a^b f(x) \, dx = G(b) = F(b) - F(a). \tag{1.86}$$

This completes the proof of the first part of the theorem.   □

We shall now present in detail the proof of the second part.

Proof. This is a limit proof by Riemann sums. Let f be (Riemann) integrable on the interval $[a, b]$, and let f admit an anti-derivative F on $[a, b]$. Begin with the quantity $F(b)F(a)$. Let there be numbers $x_1, \ldots, x_n$ such that:

$$a = x_0 < x_1 < x_2 < \cdots < x_{n-1} < x_n = b. \tag{1.87}$$

It follows that:

$$F(b) - F(a) = F(x_n) - F(x_0). \tag{1.88}$$

Now, we add each $F(x_i)$ along with its additive inverse, so that the resulting quantity is equal:

$$F(b) - F(a) = F(x_n) + [-F(x_{n-1}) + F(x_{n-1})] + \cdots + [-F(x_1) + F(x_1)] - F(x_0)$$

$$= [F(x_n) - F(x_{n-1})] + [F(x_{n-1}) + \cdots - F(x_1)] + [F(x_1) - F(x_0)]. \tag{1.89}$$

The above quantity can be written as the following sum:

$$F(b) - F(a) = \sum_{i=1}^n [F(x_i) - F(x_{i-1})]. \tag{1.90}$$

Next, we employ the mean value theorem. Stated briefly:
Let F be continuous on the closed interval $[a, b]$ and differentiable on the open interval (a, b). Then there exists some c in (a, b) such that:

$$F'(c) = \frac{F(b) - F(a)}{b - a}. \tag{1.91}$$

It follows that:

$$F'(c)(b - a) = F(b) - F(a). \tag{1.92}$$

The function F is differentiable on the interval $[a, b]$; therefore, it is also differentiable and continuous on each interval $[x_{i-1}, x_i]$. According to the mean value theorem (above):

$$F(x_i) - F(x_{i-1}) = F'(c_i)(x_i - x_{i-1}). \tag{1.93}$$

Substituting the above into Equation (1.90), we get

$$F(b) - F(a) = \sum_{i=1}^{n}[F'(c_i)(x_i - x_{i-1})]. \tag{1.94}$$

The assumption implies $F'(c_i) = f(c_i)$. Also, $x_i - x_{i-1}$ can be expressed as Δx of partition i:

$$F(b) - F(a) = \sum_{i=1}^{n}[f(c_i)(\Delta x_i)]. \tag{1.95}$$

We are describing the area of a rectangle, with the width times the height, and we are adding the areas together. Each rectangle, by virtue of the mean value theorem, describes an approximation of the curve section it is drawn over. Also Δx_i need not be the same for all values of i, or in other words that the width of the rectangles can differ. What we have to do is approximate the curve with n rectangles. Now, as the size of the partitions get smaller and n increases, resulting in more partitions to cover the space, we get closer and closer to the actual area of the curve.

By taking the limit of the expression as the norm of the partitions approaches zero, we arrive at the Riemann integral. We know that this limit exists because f was assumed to be integrable. That is, we take the limit as the largest of the partitions approaches zero in size, so that all other partitions are smaller and the number of partitions approaches infinity.

So, we take the limit on both sides of Equation (1.95). This gives us:

$$\lim_{\|\Delta x_i\| \to 0} F(b) - F(a) = \lim_{\|\Delta x_i\| \to 0} \sum_{i=1}^{n}[f(c_i)(\Delta x_i)]. \tag{1.96}$$

Neither $F(b)$ nor $F(a)$ is dependent on $\|\Delta x_i\|$, so the limit on the left side remains $F(b)F(a)$:

$$F(b) - F(a) = \lim_{\|\Delta x_i\| \to 0} \sum_{i=1}^{n}[f(c_i)(\Delta x_i)]. \tag{1.97}$$

The expression on the right side of the equation defines the integral over f from a to b. Therefore, we obtain

$$F(b) - F(a) = \int_a^b f(x)\, dx, \qquad (1.98)$$

which completes the proof.   □

It almost looks like the first part of the theorem follows directly from the second. That is, suppose G is an anti-derivative of f. Then by the second theorem:

$$G(x) - G(a) = \int_a^x f(t)\, dt. \qquad (1.99)$$

Now, suppose

$$F(x) = \int_a^x f(t)\, dt = G(x) - G(a), \qquad (1.100)$$

then F has the same derivative as G, and therefore $F = f$. This argument only works, however, if we already know that f has an anti-derivative, and the only way we know that all continuous functions have anti-derivatives is by the first part of the Fundamental Theorem. For example, if $f(x) = e^{x^2}$, then f has an anti-derivative, namely

$$G(x) = \int_0^x f(t)\, dt, \qquad (1.101)$$

and there is no simpler expression for this function. It is therefore important not to interpret the second part of the theorem as the definition of the integral. Indeed, there are many functions that are integrable but lack anti-derivatives that can be written as an elementary function. Conversely, many functions that have antiderivatives are not Riemann integrable.

1.3.2 Anti-derivative with non-integer order

Definition 1.3.2. Riemann–Liouville fractional integral [18–20]. The Riemann–Liouville fractional integral of a function f is given by the following formula:

$${}_aD_t^{-\alpha}f(t) = {}_aI_t^\alpha f(t) = \frac{1}{\Gamma(\alpha)} \int_a^t (t-\tau)^{\alpha-1} f(\tau)\, d\tau, \qquad (1.102)$$

where

$$\Gamma(t) = \int_0^\infty x^{t-1} e^{-x}\, dx. \qquad (1.103)$$

Definition 1.3.3. The Laplace transform of a function $f(t)$, defined for all real numbers $0 \leq t$, is the function $F(s)$, defined by:

$$F(s) = \int_0^\infty e^{-st} \, dt. \tag{1.104}$$

The parameter s is the complex number frequency $s = \sigma + iw$

The following of the Laplace transform properties are very useful in fractional calculus:

1.
$$\mathcal{L}\left(af(t) + bg(t)\right) = aF(s) + bF(s), \tag{1.105}$$

2.
$$\mathcal{L}\left(\int_0^t f(x)\right) = \frac{1}{s}F(s), \tag{1.106}$$

3.
$$\mathcal{L}\left(\int_0^t f(x)g(t-x) \, dx\right) = F(s)G(s). \tag{1.107}$$

Definition 1.3.2 can also be constructed using the properties of the Laplace transform method as follows. Noting that:

$$\mathcal{L}\{Jf\}(s) = \mathcal{L}\left\{\int_0^t f(\tau) \, d\tau\right\}(s) = \frac{1}{s}(\mathcal{L}\{f\})(s), \tag{1.108}$$

also, we have the following:

$$\mathcal{L}\left\{J^2 f\right\} = \frac{1}{s}(\mathcal{L}\{Jf\})(s) = \frac{1}{s^2}(\mathcal{L}\{f\})(s), \tag{1.109}$$

and so on, we assert

$$J^\alpha f = \mathcal{L}^{-1}\left\{s^{-\alpha}(\mathcal{L}\{f\})(s)\right\}. \tag{1.110}$$

For instance,

$$J^\alpha\left(t^k\right) = \mathcal{L}^{-1}\left\{\frac{\Gamma(k+1)}{s^{\alpha+k+1}}\right\} = \frac{\Gamma(k+1)}{\Gamma(\alpha+k+1)}t^{\alpha+k}, \tag{1.111}$$

as expected. Indeed, given the convolution rule:

$$\mathcal{L}\{f * g\} = (\mathcal{L}\{f\})(\mathcal{L}\{g\}), \tag{1.112}$$

and shorthanding $h(x) = x^{\alpha-1}$ for clarity, we find that:

$$(J^\alpha f)(t) = \frac{1}{\Gamma(\alpha)}\mathcal{L}^{-1}\left\{(\mathcal{L}\{p\})(\mathcal{L}\{f\})\right\}$$

$$= \frac{1}{\Gamma(\alpha)}(p * f)$$

$$= \frac{1}{\Gamma(\alpha)} \int_0^t p(t - \tau) f(\tau) \, d\tau$$

$$= \frac{1}{\Gamma(\alpha)} \int_0^t (t - \tau)^{\alpha - 1} f(\tau) \, d\tau. \qquad (1.113)$$

The Hadamard fractional integral is introduced by J. Hadamard and is given by the following formula:

$$_a D_t^{-\alpha} f(t) = \frac{1}{\Gamma(\alpha)} \int_a^t \left(\log \frac{t}{\tau} \right)^{\alpha - 1} f(\tau) \frac{d\tau}{\tau}, \quad t > a. \qquad (1.114)$$

Erdlyi–Kober operator. The Erdlyi–Kober operator is an integral operator introduced by Arthur Erdlyi (1940) and Hermann Kober (1940), and is given by:

$$\frac{x^{-\nu - \alpha + 1}}{\Gamma(\alpha)} \int_0^x (t - x)^{\alpha - 1} t^{-\alpha - \nu} f(t) \, dt, \qquad (1.115)$$

which generalizes the Riemann fractional integral and the Weyl integral. A recent generalization is the following, which generalizes the Riemann–Liouville fractional integral and the Hadamard fractional integral. It is given by:

$$\left({}^\rho \mathcal{I}_{a+}^\alpha f \right)(x) = \frac{\rho^{1 - \alpha}}{\Gamma(\alpha)} \int_a^x \frac{\tau^{\rho - 1} f(\tau)}{(x^\rho - \tau^\rho)^{1 - \alpha}} \, d\tau, \quad x > a. \qquad (1.116)$$

It is straightforward to show that the J operator satisfies:

$$(J^\alpha)(J^\beta f)(x) = (J^\beta)(J^\alpha f)(x) = (J^{\alpha + \beta} f)(x)$$

$$= \frac{1}{\Gamma(\alpha + \beta)} \int_0^x (x - t)^{\alpha + \beta - 1} f(t) \, dt, \qquad (1.117)$$

Proof.

$$(J^\alpha)(J^\beta f)(x) = \frac{1}{\Gamma(\alpha)} \int_0^x (x - t)^{\alpha - 1} (J^\beta f)(t) \, dt$$

$$= \frac{1}{\Gamma(\alpha)\Gamma(\beta)} \int_0^x \int_0^t (x - t)^{\alpha - 1} (t - s)^{\beta - 1} f(s) \, ds \, dt$$

$$= \frac{1}{\Gamma(\alpha)\Gamma(\beta)} \int_0^x f(s) \left(\int_s^x (x-t)^{\alpha-1}(t-s)^{\beta-1} \, dt \right) ds.$$

$$(1.118)$$

where in the last step we exchanged the order of integration and pulled out the $f(s)$ factor from the t integration. Changing variables to r defined by $t = s + (xs)r$:

$$(J^\alpha)(J^\beta f)(x) = \frac{1}{\Gamma(\alpha)\Gamma(\beta)} \int_0^x (x-s)^{\alpha+\beta-1}f(s) \left(\int_0^1 (1-r)^{\alpha-1}r^{\beta-1} \, dr \right) ds.$$

$$(1.119)$$

The inner integral is the beta function which satisfies the following property:

$$\int_0^1 (1-r)^{\alpha-1}r^{\beta-1} \, dr = B(\alpha, \beta) = \frac{\Gamma(\alpha)\Gamma(\beta)}{\Gamma(\alpha+\beta)}. \qquad (1.120)$$

Substituting back into the equation:

$$(J^\alpha)(J^\beta f)(x) = \frac{1}{\Gamma(\alpha+\beta)} \int_0^x (x-s)^{\alpha+\beta-1}f(s) \, ds = (J^{\alpha+\beta}f)(x). \quad (1.121)$$

Interchanging α and β shows that the order in which the J operator is applied is irrelevant and completes the proof. $\square$

1.3.3 Integral of variable order

VO fractional integral, which is an extension of constant, has been introduced in several physical problems [32–36]. The VO fractional derivative is good in depicting memory properties that change with time or space location. Let us recall the relevant definitions for VO fractional calculus.

Definition 1.3.4. Left and right Riemann–Liouville integrals of variable order: Let $0 < \alpha(x, t) < 0$ for all $(x, t) \in [a, b]$ and $f \in \mathbb{L}_{\mathcal{K}}$ then

$$_aI_t^{\alpha(.)(f(t))} = \int_a^t \frac{1}{\Gamma[\alpha(t,x)]}(t-x)^{\alpha(t,x)-1}f(x) \, dx, \quad (t > a) \qquad (1.122)$$

is called the left Riemann–Liouville integral of variable fractional order $\alpha(.,.)$ and

$$_bI_t^{\alpha(.)(f(t))} = \int_t^b \frac{1}{\Gamma[\alpha(t,x)]}(x-t)^{\alpha(t,x)-1}f(x) \, dx, \quad (b < t) \qquad (1.123)$$

is referred to as the right Riemann–Liouville integral of variable fractional order $\alpha(.,.)$.

1.4 LIMITATIONS AND STRENGTH OF LOCAL AND FRACTIONAL DERIVATIVES

Fractional calculus has been used to model physical and engineering processes, which are found to be best described by fractional differential equations. It is worth noting that the standard mathematical models of integer-order derivatives, including nonlinear models, do not work adequately in many cases. In recent years, fractional calculus has played a very important role in various fields such as mechanics, electricity, chemistry, biology, economics, notably control theory, and signal and image processing. Major topics include anomalous diffusion, vibration and control, continuous time random walk, Levy statistics, fractional Brownian motion, fractional neutron point kinetic model, power law, Riesz potential, fractional derivative and fractals, computational fractional derivative equations, nonlocal phenomena, history-dependent process, porous media, fractional filters, biomedical engineering, fractional phase-locked loops, fractional variational principles, fractional transforms, fractional wavelet, fractional predator-prey system, soft matter mechanics, fractional signal and image processing; singularities analysis and integral representations for fractional differential systems; special functions related to fractional calculus, non-Fourier heat conduction, acoustic dissipation, geophysics, relaxation, creep, viscoelasticity, rheology, fluid dynamics, chaos, and groundwater problems. An excellent literature for all this can be found [37–47]. It is very important to point out that all these fractional derivative order definitions have their advantages and disadvantages; here we shall include Caputo, variational order, Riemann–Liouville Jumarie, and Weyl. We shall examine first the variational order differential operator.

1.4.1 Advantages of fractional derivatives

1. Anomalous diffusion phenomena are extensively observed in physics, chemistry, and biology fields [18–21]. To characterize anomalous diffusion phenomena, constant-order fractional diffusion equations are introduced and have received tremendous success. However, it has been found that the constant-order fractional diffusion equations are not capable of characterizing some complex diffusion processes, for instance, diffusion process in inhomogeneous or heterogeneous medium [22]. In addition, when we consider diffusion process in porous medium, if the medium structure or external field changes with time, in this situation, the constant-order fractional diffusion equation model cannot be used to characterize such phenomena well [23, 24]. Still in some

biology diffusion processes, the concentration of particles will determine the diffusion pattern [25, 26]. To solve the above problems, the VO fractional diffusion equation models have been suggested for use [27]. The ground-breaking work of VO operator can be traced to Samko et al. by introducing the VO integration and Riemann–Liouville derivative in [27]. It has been recognized as a powerful modeling approach in the fields of viscoelasticity [17–32], viscoelastic deformation [28], viscous fluid [29], and anomalous diffusion [30].

2. With the Jumarie definition, which is actually the modified Riemann–Liouville fractional derivative, an arbitrary continuous function needs not to be differentiable; the fractional derivative of a constant is equal to zero and more importantly it removes singularity at the origin for all functions for which, for instance, the exponentials functions and Mittag–Leffler functions.

3. With the Riemann–Liouville fractional derivative, an arbitrary function needs not to be continuous at the origin and it needs not to be differentiable.

4. One of the great advantages of the Caputo fractional derivative is that it allows traditional initial and boundary conditions to be included in the formulation of the problem [4, 19, 37, 38]. In addition, its derivative for a constant is zero. It is customary in groundwater investigations to choose a point on the centerline of the pumped borehole as a reference for the observations and therefore neither the drawdown nor its derivatives will vanish at the origin, as required [48]. In such situations where the distribution of the piezometric head in the aquifer is a decreasing function of the distance from the borehole, the problem may be circumvented by rather using the complementary, or Weyl, fractional order derivative [33]. The Caputo fractional derivative also allows the use of the initial and boundary conditions when dealing with real-world problems. The Caputo derivative is the most appropriate fractional operator to be used in modeling real-world problem.

5. q-analogs find applications in a number of areas, including the study of fractals and multifractal measures, and expressions for the entropy of chaotic dynamical systems. The relationship to fractals and dynamical systems results from the fact that many fractal patterns have the symmetries of Fuchsian groups in general (e.g., Indra's pearls and the Apollonian gasket) and the modular group in particular. The connection passes through hyperbolic geometry and ergodic theory, where the elliptic integrals and modular forms play a prominent role; the q-series themselves are closely related to elliptic integrals.

6. The local fractional derivatives, being local in nature, have proven useful in studying fractional differentiability properties of highly irregular and nowhere differentiable functions. The derivative obeys Leibniz rule and chain rule.

7. As an alternative modeling approach to the classical Ficks second law, the fractal derivative is used to derive a linear anomalous transport-diffusion equation underlying anomalous diffusion process.

1.4.2 Disadvantages of fractional derivatives

Although these fractional derivatives display great advantages, they are not applicable in all the situations. We shall begin with the Riemann–Liouville type.

1. The Riemann–Liouville derivative has certain disadvantages when trying to model real-world phenomena with fractional differential equations. The Riemann–Liouville derivative of a constant is not zero. In addition, if an arbitrary function is a constant at the origin, its fractional derivation has a singularity at the origin for instant exponential and Mittag–Leffler functions. Theses disadvantages reduce the field of application of the Riemann–Liouville fractional derivative.

2. Caputo's derivative demands higher conditions of regularity for differentiability: to compute the fractional derivative of a function in the Caputo sense, we must first calculate its derivative. Caputo derivatives are defined only for differentiable functions while functions that have no first-order derivative might have fractional derivatives of all orders less than one in the Riemann–Liouville sense.

3. With the Jumarie fractional derivative, if the function is not continuous at the origin, the fractional derivative will not exist, for instance, what will be the fractional derivative of $\ln(x)$ and many other functions.

4. Although the Weyl fractional derivative found its place in groundwater investigation, it still displays a significant disadvantage; because the integral defining these Weyl derivatives is improper, greater restrictions must be placed on a function. For instance, the Weyl derivative of a constant is not defined. On the other hand, general theorem about Weyl derivatives are often more difficult to formulate and be proved than are corresponding theorems for Riemann–Liouville derivatives.

5. The fractional VO derivatives do not obey basic properties of derivative and they are also very difficult, not to say impossible, to handle analytically. For instance, we do not know so far what is the VO fractional derivative of the function x. Differential equation, evolving

these derivatives are not possible to solve with any analytical method. In general, variational order differential operator cannot easily be handled analytically. A numerical approach is sometimes needed to deal with the problem under investigation.

1.5 CLASSIFICATION OF FRACTIONAL DERIVATIVES

Because of the usefulness of the concept of derivative, many researchers have paid attention to this concept. Therefore, there exist many operators in the literature that are called derivative. The real questions that must be asked at this stage are: "What is the derivative? Which operator can be called fractional derivative?" This section discusses the concepts underlying the formulation of operators capable of being interpreted as fractional derivatives. I have found in the literature a paper [49] that attempted to provide some criteria that must be satisfied by an operator to be called fractional derivative. However, in that paper, the authors provide criteria that must be obeyed by an operator to be able to be called derivative. We shall mention that any fractional derivative must satisfy some basic properties of derivatives before being fractional.

1.5.1 Criteria of fractional derivatives
In this section, we shall provide criteria that need to be satisfied for a given operator to be called fractional derivative. Let G be an operator, then G is called a fractional derivative if the following apply:

1. G of a function at a chosen input value describes the rate of change of the function near that input value.
2. $G(f)$ of order zero produces f.
3. G is linear, that is:

$$G(af(t) + bg(t)) = aG(f) + bG(t). \tag{1.218}$$

4. G satisfies the constant factor rule, that is:

$$G(af) = aG(f). \tag{1.219}$$

5. G satisfies the sum rule, that is:

$$G(f+g) = G(f) + G(g). \tag{1.220}$$

6. G satisfies the subtraction rule, that is:

$$G(f-g) = G(f) - G(g). \tag{1.221}$$

7. G satisfies the product rule, that is for the functions f and g, G of the function $h(x) = f(x)g(x)$ with respect to x is:

$$G(h(x)) = G(f(x))g(x) + f(x)G(g(x)). \qquad (1.222)$$

8. G satisfies the chain rule, that is, for the functions f and g, G of the function $h(x) = f(g(x))$ with respect to x is:

$$G(h(x)) = G(f(x))g'(x) = f'(x)G(g(x)). \qquad (1.223)$$

9. G satisfies the inverse function, that is, if the function f has an inverse function g, meaning that $g(f(x)) = x$ and $f(g(y)) = y$, then

$$G(g) = \frac{l(t, \beta)}{G(f(g))}. \qquad (1.224)$$

10. G satisfies the reciprocal rule, G of $h(x) = 1/f(x)$ for any (nonvanishing) function f is:

$$G(h(x)) = -\frac{G(f(x))}{(f(x))^2}, \qquad (1.225)$$

11. G satisfies the quotient rule, that is, if f and g are functions, then

$$G\left(\frac{f}{g}\right) = \frac{G(f)g - G(g)f}{g^2}. \qquad (1.226)$$

   wherever g is nonzero:
12. $G(f(x))$ is positive in an open interval, then f is increasing in that interval.
13. $G(f(x))$ is negative in an open interval, then f is decreasing in that interval.
14. $G(f(x)) = 0$ in an open interval, then f is constant in that interval.
15. G satisfies the above criteria and D its inverse operator, then D is a fractional integral.
16. G satisfies the backward compatibility, meaning, when the order is integer, G gives the same results as ordinary derivatives.

We shall use the above criteria to classify those operators found in the literature. We shall start with the commonly used fractional operator (Caputo and Riemann–Liouville). The Caputo fractional operator does not satisfy:

1. G of a function at a chosen input value describes the rate of change of the function near that input value.
2. G satisfies the chain rule, that is for the functions f and g, G of the function $h(x) = f(g(x))$ with respect to x is:

$$G(h(x)) = G(f(x))g'(x) = f'(x)G(g(x)). \qquad (1.227)$$

3. G satisfies the inverse function, that is, if the function f has an inverse function g, meaning that $g(f(x)) = x$ and $f(g(y)) = y$, then

$$G(g) = \frac{l(t, \beta)}{G(f(g))}. \tag{1.228}$$

4. G satisfies the reciprocal rule, G of $h(x) = 1/f(x)$ for any (nonvanishing) function f is:

$$G(h(x)) = -\frac{G(f(x))}{(f(x))^2}. \tag{1.229}$$

The Riemann–Liouville fractional operator does not satisfy the following:

1. G of a function at a chosen input value describes the rate of change of the function near that input value.
2. G satisfies the chain rule, that is for the functions f and g, G of the function $h(x) = f(g(x))$ with respect to x is:

$$G(h(x)) = G(f(x))g'(x) = f'(x)G(g(x)). \tag{1.230}$$

3. G satisfies the inverse function, that is, if the function f has an inverse function g, meaning that $g(f(x)) = x$ and $f(g(y)) = y$, then

$$G(g) = \frac{l(t, \beta)}{G(f(g))}. \tag{1.231}$$

4. G satisfies the reciprocal rule, G of $h(x) = 1/f(x)$ for any (nonvanishing) function f is:

$$G(h(x)) = -\frac{G(f(x))}{(f(x))^2}. \tag{1.232}$$

5. G satisfies the backward compatibility, meaning when the order is integer, G gives the same results as ordinary derivatives.

We can therefore conclude that both the Riemann–Liouville and Caputo operators are not derivatives, and then they are not fractional derivatives, but fractional operators. We agree with the result [30] that, the local fractional operator is not a fractional derivative. One of the aims of this book is to provide a suitable derivative with fractional order derivative, that satisfies criteria (1–16), and this is presented in the next chapter.

CHAPTER 2

Local derivative with new parameter

2.1 MOTIVATION

A large class of singular perturbed problems, the domain may be divided into two or more sub-domains. In one of these, often the largest, the solution is accurately approximated by an asymptotic series found by treating it as a regular perturbation. The other sub-domains consist of one or more small areas in which that approximation is inaccurate, generally because the perturbation terms in the problem are not negligible there. These areas are referred to as transition layers, and as boundary or interior layers depending on whether they occur at the domain boundary or inside the domain [50–55]. An approximation in the form of an asymptotic series is obtained in transition layers by treating that part of the domain as separate from the domain as a separate perturbation problem. This approximation is called the inner solution, and the other is the outer solution named for their relationship to transition layers. The outer and inner solutions are then combined through a process called matching in such a way that an approximate solution for the whole domain is obtained [50–55]. Therefore, the method of matched asymptotic is a common approach to finding an accurate approximation to solution to an equation in particular when solving perturbed differential equation with conventional order derivatives. This class of differential equations is used to describe real-world problems, for instance, in physic and fluid mechanics, a boundary layer is the layer of fluid in the immediate neighborhood of a bouncing surface anywhere; the consequences of viscidness are noteworthy. In the Earth's atmosphere, the environmental borderline layer is the air layer neighboring the earth affected by diurnal heat, moisture, or momentum transfer to or from the surface. On an aircraft wing the boundary layer is the part of the flow close to the wing, where viscous forces distort the surrounding nonviscous flow. In recent decades, attention has been paid by several scholars in modeling the real-world problems with the concept of fractional order derivatives. It was revealed with many proofs and results that the modeling of these real-world problems with the concept of fractional order derivatives gives better predictions rather than using conventional derivatives, which are regarded as integer order derivatives. It is perhaps important to mention

Derivative with a New Parameter. http://dx.doi.org/10.1016/B978-0-08-100644-3.00002-7
© 2016 Elsevier Ltd. All rights reserved.

that the match asymptotic method have never been used to solve any kind of fractional differential equations, because of the nature and properties of the fractional derivative. In particular, the most commonly used fractional derivative (Caputo's fractional derivative) for modeling real-world problems does not obey the chain rule which is one of the key elements of the match asymptotic method. Recently, the so-called conformable fractional derivative was proposed. This fractional derivative is theoretically very easy to handle and also obeys some conventional properties that cannot be satisfied by the existing fractional derivatives, for instance, the chain rule. However, this fractional derivative has a very big weakness, which is the fractional derivative of any differentiable function at the point zero is zero and this does not satisfy any physical problem or cannot for the moment have any physical interpretation. A modified version was proposed in order to extend the limitation of the conformable fractional derivative [56–60]; however, this derivative depends on the interval on which the function is being fractional differentiated, which is also a true problem for some physical problems. The aim of this new derivative is to extend further the well-known match asymptotic method in the scope of the fractional differential equation, to be used to describe further the boundary layers problems within the folder of fractional calculus on one hand. On the other hand, the derivative with new parameter will obey the 16 criteria in Section 1.5.

2.2 DEFINITION AND ANTI-DERIVATIVE

Definition 2.2.1. Let $a \in \mathbb{R}$ and g be a function, such that, $g: [a, \infty) \to \mathbb{R}$. Then, the β-derivative of g is defined as:

$$
{}_0^A D_t^\beta g(t) =
\begin{cases}
\lim\limits_{\varepsilon \to 0} \dfrac{g\left(t + \varepsilon\left(t + \frac{1}{\Gamma(\beta)}\right)^{1-\beta}\right) - g(t)}{\varepsilon} & \text{for all } t \geq 0,\ 0 < \beta \leq 1 \\
g(t) & \text{for all } t \geq 0,\ \beta = 0,
\end{cases}
$$

$$\tag{2.1}$$

where g is a function such that $g: [0, \infty) \to \mathbb{R}$ and Γ the gamma-function

$$
\Gamma(\zeta) = \int_0^\infty t^{\zeta - 1}\, e^{-1}\, dt.
$$

If the above limit of exists then g is said to be β-differentiable. Note that for $\beta = 1$, we have ${}_0^A D_t^\beta g(t) = \frac{d}{dt} g(t)$. Moreover, unlike other fractional derivatives, the β-derivative of a function can be locally defined at a certain point, the same way like first-order derivative.

Theorem 2.2.1. *The local derivative with new parameter satisfies criteria (1–16) of Section 1.5.*

Proof. Let f and g be differential and beta-differentiable on $[a, b]$, let μ and α be two real number, then

$$\,^{A}_{0}D^{0}_{t}g(t) = g(t). \tag{2.2}$$

This is by definition, then number 2 is satisfied. $\qquad\square$

We shall check the third criteria 3, 4, 5, and 6.

Proof. By definition, we have

$$\,^{A}_{0}D^{\beta}_{t}(\alpha g(t) + \mu f(t)) = \lim_{\varepsilon \to 0} \frac{(\alpha g + \mu f)\left(t + \varepsilon \left(t + \frac{1}{\Gamma(\beta)}\right)^{1-\beta}\right) - (\alpha g + \mu f)(t)}{\varepsilon}. \tag{2.3}$$

Rearranging, we obtain

$$\lim_{\varepsilon \to 0} \frac{(\alpha g + \mu f)\left(t + \varepsilon \left(t + \frac{1}{\Gamma(\beta)}\right)^{1-\beta}\right) - (\alpha g + \mu f)(t)}{\varepsilon}$$

$$= \alpha \lim_{\varepsilon \to 0} \frac{g\left(t + \varepsilon \left(t + \frac{1}{\Gamma(\beta)}\right)^{1-\beta}\right) - g(t)}{\varepsilon}$$

$$+ \mu \lim_{\varepsilon \to 0} \frac{f\left(t + \varepsilon \left(t + \frac{1}{\Gamma(\beta)}\right)^{1-\beta}\right) - f(t)}{\varepsilon}, \tag{2.4}$$

and finally we obtain the requested result as follows:

$$\,^{A}_{0}D^{\beta}_{t}(\alpha g(t) + \mu f(t)) = \alpha\,^{A}_{0}D^{\beta}_{t}(g(t)) + \mu\,^{A}_{0}D^{\beta}_{t}(f(t)). \tag{2.5}$$

Note that $\alpha = 0, 1, -1$ we cover criteria 4, 5, and 6, and this completed the proof. $\qquad\square$

We shall present the proof of 7.

Proof. By definition, we have

$$
{}_0^A D_t^\beta (g(t)f(t)) = \lim_{\varepsilon \to 0} \frac{(gf)\left(t+\varepsilon\left(t+\frac{1}{\Gamma(\beta)}\right)^{1-\beta}\right) - (gf)(t)}{\varepsilon}. \tag{2.6}
$$

However, the numerator of Equation (2.6) can be reformulated as follows:

$$
(gf)\left(t+\varepsilon\left(t+\frac{1}{\Gamma(\beta)}\right)^{1-\beta}\right) - (gf)(t) = (gf)\left(t+\varepsilon\left(t+\frac{1}{\Gamma(\beta)}\right)^{1-\beta}\right) - (gf)(t)
$$

$$
+ f\left(t+\varepsilon\left(t+\frac{1}{\Gamma(\beta)}\right)^{1-\beta}\right) g(t)
$$

$$
- f\left(t+\varepsilon\left(t+\frac{1}{\Gamma(\beta)}\right)^{1-\beta}\right) g(t). \tag{2.7}
$$

Now replacing Equation (2.7) into Equation (2.6) and rearranging, we obtain the following:

$$
{}_0^A D_t^\beta (g(t)f(t)) = \lim_{\varepsilon \to 0} \frac{\left[f\left(t+\varepsilon\left(t+\frac{1}{\Gamma(\beta)}\right)^{1-\beta}\right) - f(t)\right] g(t)}{\varepsilon}
$$

$$
+ \lim_{\varepsilon \to 0} \frac{f\left(t+\varepsilon\left(t+\frac{1}{\Gamma(\beta)}\right)^{1-\beta}\right)\left[g\left(t+\varepsilon\left(t+\frac{1}{\Gamma(\beta)}\right)^{1-\beta}\right) - g(t)\right]}{\varepsilon}. \tag{2.8}
$$

Using the properties of the limit for a continuous function, we obtain the requested result:

$$
{}_0^A D_t^\beta (g(t)f(t)) = {}_0^A D_t^\beta (g(t))f(t) + {}_0^A D_t^\beta (g(t))f(t). \tag{2.9}
$$

This completes the proof of criteria 7. We shall now present the proof of criteria 8. □

Proof. By definition, we have

$$
{}_0^A D_t^\beta ((g \circ f)(t)) = \lim_{\varepsilon \to 0} \frac{(g \circ f)\left(t+\varepsilon\left(t+\frac{1}{\Gamma(\beta)}\right)^{1-\beta}\right) - (g \circ f)(t)}{\varepsilon}. \tag{2.10}
$$

For simplicity, let $h = \varepsilon \left(t + \frac{1}{\Gamma(\beta)}\right)^{1-\beta}$. Replacing this in Equation (2.12), we obtain

$$_0^A D_t^\beta((g \circ f)(t)) = \left(t + \frac{1}{\Gamma(\beta)}\right)^{1-\beta} \lim_{h \to 0} \frac{(g \circ f)(t+h) - (g \circ f)(t)}{h}. \quad (2.11)$$

However, using the fact that f and g are differentiable together with chain rule, we obtain the following:

$$_0^A D_t^\beta((g \circ f)(t)) = \left(t + \frac{1}{\Gamma(\beta)}\right)^{1-\beta} \lim_{h \to 0} \frac{g(f(t)+h) - g(f(t))}{h} \lim_{h \to 0} \frac{f(h+t) - f(t)}{h}. \quad (2.12)$$

Since f is differentiable, we have

$$f'(t) = \lim_{h \to 0} \frac{f(t+h) - f(t)}{h}. \quad (2.13)$$

Now replacing h by $\varepsilon \left(t + \frac{1}{\Gamma(\beta)}\right)^{1-\beta}$, we obtain

$$_0^A D_t^\beta((g \circ f)(t)) = \lim_{\varepsilon \to 0} \frac{g\left(f(t) + \varepsilon \left(t + \frac{1}{\Gamma(\beta)}\right)^{1-\beta}\right) - g(f(t))}{\varepsilon} f', \quad (2.14)$$

therefore, by definition, we obtain the requested formula:

$$_0^A D_t^\beta((g \circ f)(t)) = {_0^A D_t^\beta}(g(f(t)))f'(x). \quad (2.15)$$

This completes the proof of criteria 8. We shall present the detail proof of criteria 9. □

Proof. Assume that the function f has an inverse function g, meaning that $g(f(x)) = x$ and $f(g(y)) = y$, then

$$_0^A D_t^\beta((g \circ f)(t)) = {_0^A D_t^\beta}(t) = \left(t + \frac{1}{\Gamma(\beta)}\right)^{1-\beta}. \quad (2.16)$$

Now using criteria 8, Equation (2.16) becomes:

$$_0^A D_t^\beta f(t) = \frac{l(t,\beta)}{_0^A D_t^\beta [g(f(t))]}, \quad l(t,\beta) = \left(t + \frac{1}{\Gamma(\beta)}\right)^{2-2\beta}. \quad (2.17)$$

This produces the requested result. We shall present the detail proof of criteria 10. □

Proof. By definition, we have the following formula:

$$
{}_0^A D_t^\beta (f^{-1}(t)) = \lim_{\varepsilon \to 0} \frac{f^{-1}\left(t + \varepsilon \left(t + \frac{1}{\Gamma(\beta)}\right)^{1-\beta}\right) - f^{-1}(t)}{\varepsilon}
$$

$$
= -\lim_{\varepsilon \to 0} \frac{f\left(t + \varepsilon \left(t + \frac{1}{\Gamma(\beta)}\right)^{1-\beta}\right) - f(t)}{\varepsilon f\left(t + \varepsilon \left(t + \frac{1}{\Gamma(\beta)}\right)^{1-\beta}\right) f(t)}
$$

$$
= -\frac{{}_0^A D_t^\beta (f(t))}{f^2(x)}. \tag{2.18}
$$

This produces the requested formula, then criteria 10 is satisfied. We shall present next the detail proof of 11.   □

Proof. By definition, we have the following equation:

$$
{}_0^A D_t^\beta \left(\frac{f(t)}{g(t)}\right) = \lim_{\varepsilon \to 0} \frac{(f \cdot g^{-1})\left(t + \varepsilon \left(t + \frac{1}{\Gamma(\beta)}\right)^{1-\beta}\right) - (f \cdot g^{-1})(t)}{\varepsilon}. \tag{2.19}
$$

However, making use of criteria 7 and 10 and rearranging, we obtain

$$
{}_0^A D_t^\beta \left(\frac{f(t)}{g(t)}\right) = \frac{{}_0^A D_t^\beta (g(t))f(t) - {}_0^A D_t^\beta (g(t))f(t)}{g^2(t)}. \tag{2.20}
$$

This completes the proof of criteria 11. We shall now present the detail proof of criteria 12.   □

Proof. We assume that ${}_0^A D_t^\beta (f(t))$ is positive in an open interval, then f is increasing in that interval, then using the definition of the derivative with new parameter, we obtain and considering $t_1 > t_2$:

$$
\frac{f\left(t_1 + \varepsilon \left(t_2 + \frac{1}{\Gamma(\beta)}\right)^{1-\beta}\right) - f(t_2)}{\varepsilon} > 0. \tag{2.21}
$$

Then

$$
f\left(t_1 + \varepsilon \left(t_2 + \frac{1}{\Gamma(\beta)}\right)^{1-\beta}\right) - f(t_2) > 0. \tag{2.22}
$$

Now, taking the limit on both sides of Equation (2.22), we have

$$\lim_{\varepsilon \to 0} f\left(t_1 + \varepsilon \left(t_2 + \frac{1}{\Gamma(\beta)}\right)^{1-\beta}\right) - f(t_2) = f(t_1) - f(t_2) > 0. \qquad (2.23)$$

This completes the proof of criteria 12. We shall show now the proof of criteria 13.   □

Proof. We assume that $_0^A D_t^\beta (f(t))$ is negative in an open interval, then f is decreasing in that interval, then using the definition of the derivative with new parameter, we obtain and considering $t_1 > t_2$:

$$\frac{f\left(t_1 + \varepsilon \left(t_2 + \frac{1}{\Gamma(\beta)}\right)^{1-\beta}\right) - f(t_2)}{\varepsilon} < 0. \qquad (2.24)$$

Then

$$f\left(t_1 + \varepsilon \left(t_2 + \frac{1}{\Gamma(\beta)}\right)^{1-\beta}\right) - f(t_2) < 0. \qquad (2.25)$$

Now, taking the limit on both sides of Equation (2.25), we have

$$\lim_{\varepsilon \to 0} f\left(t_1 + \varepsilon \left(t_2 + \frac{1}{\Gamma(\beta)}\right)^{1-\beta}\right) - f(t_2) = f(t_1) - f(t_2) < 0. \qquad (2.26)$$

This completes the proof of criteria 13. We shall now show the proof of criteria 14.   □

Proof. We assume that $_0^A D_t^\beta (f(t))$ is zero in an open interval, then f is constant in that interval, then using the definition of the derivative with new parameter, we obtain and considering $t_1 > t_2$:

$$\frac{f\left(t_1 + \varepsilon \left(t_2 + \frac{1}{\Gamma(\beta)}\right)^{1-\beta}\right) - f(t_2)}{\varepsilon} = 0. \qquad (2.27)$$

Then

$$f\left(t_1 + \varepsilon \left(t_2 + \frac{1}{\Gamma(\beta)}\right)^{1-\beta}\right) - f(t_2) = 0. \qquad (2.28)$$

Now, taking the limit on both sides of Equation (2.28), we have

$$\lim_{\varepsilon \to 0} f\left(t_1 + \varepsilon \left(t_2 + \frac{1}{\Gamma(\beta)}\right)^{1-\beta}\right) - f(t_2) = f(t_1) - f(t_2) = 0. \quad (2.29)$$

This completes the proof of criteria 14. And finally criteria 16 is satisfied.

□

2.3 PROPERTIES OF LOCAL DERIVATIVE WITH NEW PARAMETER

We present in this section some useful theorems and properties of the derivative with new parameter.

Theorem 2.3.1. *Assume that a given function says f: $[a, \infty] \to \mathbb{R}$ is β-differentiable at a point says $x_0 \geq a$, $\beta \in [0, 1]$, then f is continuous at x_0.*

Proof. Assume that f is β-differentiable, then

$$_0^A D_t^\beta (f(t_0)) = \lim_{\varepsilon \to 0} \frac{f\left(t_0 + \varepsilon \left(t_0 + \frac{1}{\Gamma(\beta)}\right)^{1-\beta}\right) - f(t_0)}{\varepsilon} \quad (2.30)$$

exists. However,

$$\lim_{\varepsilon \to 0} f\left(t_0 + \varepsilon \left(t_0 + \frac{1}{\Gamma(\beta)}\right)^{1-\beta}\right) - f(t_0) = \lim_{\varepsilon \to 0} \frac{f\left(t_0 + \varepsilon \left(t_0 + \frac{1}{\Gamma(\beta)}\right)^{1-\beta}\right) - f(t_0)}{\varepsilon} \varepsilon$$

$$= _0^A D_f^\beta (f(t_0)).0$$

$$= 0, \quad (2.31)$$

therefore,

$$\lim_{\varepsilon \to 0} f\left(t_0 + \varepsilon \left(t_0 + \frac{1}{\Gamma(\beta)}\right)^{1-\beta}\right) = \left(\lim_{\varepsilon \to 0} f\left(t_0 + \varepsilon \left(t_0 + \frac{1}{\Gamma(\beta)}\right)^{1-\beta}\right) - f(t_0)\right) + f(t_0)$$

$$= 0 + f(t_0)$$

$$= f(t_0). \quad (2.32)$$

Nevertheless, if we assume that, $t = t_0 + \varepsilon \left(t_0 + \frac{1}{\Gamma(\beta)}\right)^{1-\beta}$ so that,

$$\varepsilon = \frac{(t - t_0)}{\left(t_0 + \frac{1}{\Gamma(\beta)}\right)^{1-\beta}}, \quad (2.33)$$

thus,

$$\lim_{\frac{(t-t_0)}{\left(t_0+\frac{1}{\Gamma(\beta)}\right)^{1-\beta}}\to 0} f(t) = f(t_0), \tag{2.34}$$

since $\left(t_0 + \frac{1}{\Gamma(\beta)}\right)^{1-\beta} \neq 0$, then Equation (2.34) can be rewritten as:

$$\lim_{t-t_0\to 0} f(t) = f(t_0) \quad \text{or} \quad \lim_{t\to t_0} f(t) = f(t_0). \tag{2.35}$$

This completes the proof. □

Theorem 2.3.2. *Assume that a given function says f: $[a, \infty] \to \mathbb{R}$ is locally differentiable then, f is also β-differentiable.*

Proof. If f is differentiable then, the following limit exists:

$$\lim_{h\to 0} \frac{f(t+h) - f(t)}{h}. \tag{2.36}$$

This implies the existence of the following limit too:

$$\lim_{h\to 0} \frac{f(t+h) - f(t)}{h} \left(t + \frac{1}{\Gamma(\beta)}\right)^{1-\beta}. \tag{2.37}$$

Nevertheless, letting $\varepsilon = h\left(t + \frac{1}{\Gamma(\beta)}\right)^{\beta-1}$ that will imply $h = \varepsilon\left(t + \frac{1}{\Gamma(\beta)}\right)^{1-\beta}$, thus Equation (2.37) can be reformulated as:

$$\lim_{\varepsilon\to 0} \frac{f\left(t + \varepsilon\left(t + \frac{1}{\Gamma(\beta)}\right)^{1-\beta}\right) - f(t)}{\varepsilon} = {}_{0}^{A}D_{t}^{\beta}(f(t)). \tag{2.38}$$

This completes the proof. □

Theorem 2.3.3. *Formal statement of the mean value theorem for variable order derivative: Let f: $[a, b] \to \mathbb{R}$ be a continuous function on the closed interval $[a, b]$, and β-differentiable and differentiable on the open interval (a, b), where $a < b$. Then there exists some c in (a, b) such that:*

$$_{0}^{A}D_{t}^{\beta}(f(c)) = h(\beta, c, a, b)\frac{f(b) - f(a)}{b - a}. \tag{2.39}$$

Proof. According to the basic idea of the local derivative, the sloop that joins the points $(a, f(a))$ and $(b, f(b))$ is given by the expression $\frac{f(b)-f(a)}{b-a}$ [61]. This is in other words the chord of the graph of the function f, which in physical or geometrical interpretation gives the $f'(x)$ slope of the tangent to the curve at the point $(t, f(t))$. Let us define

$$J(t) = f(t) - d\left(t + \frac{1}{\Gamma(\beta)}\right)^{1-\beta} t, \tag{2.40}$$

where d and c are constants. Thus f is continuous on $[a, b]$ and differentiable on (a, b), therefore J is differentiable. However, we choose d such that the Rolle's theorem can be satisfied including:

$$J(a) = J(b) \equiv f(a) - d\left(a + \frac{1}{\Gamma(\beta)}\right)^{1-\beta} a = f(b) - d\left(b + \frac{1}{\Gamma(\beta)}\right)^{1-\beta} b. \tag{2.41}$$

However, rearranging, we obtain the following:

$$d = \frac{f(b) - f(a)}{b\left(b + \frac{1}{\Gamma(\beta)}\right)^{1-\beta} - a\left(a + \frac{1}{\Gamma(\beta)}\right)^{1-\beta}}. \tag{2.42}$$

Thus by Rolle's theorem, since J is differentiable and $J(a) = J(b)$, we can then find a constant $c \in (a, b)$ such that $J'(c) = 0$. However,

$$J'(x) = f'(x) - d\left(t + \frac{1}{\Gamma(\beta)}\right)^{1-\beta} - d(1-\beta)\left(t + \frac{1}{\Gamma(\beta)}\right)^{-\beta}. \tag{2.43}$$

Thus, $J'(c) = 0$ implies

$$d = \frac{f'(c)}{\left(c + \frac{1}{\Gamma(\beta)}\right)^{1-\beta} + \left(c + \frac{1}{\Gamma(\beta)}\right)^{-\beta}}. \tag{2.44}$$

Nevertheless, comparing Equations (2.42) and (2.44), we obtain

$$\frac{f'(c)}{\left(c + \frac{1}{\Gamma(\beta)}\right)^{1-\beta} + \left(c + \frac{1}{\Gamma(\beta)}\right)^{-\beta}} = \frac{f(b) - f(a)}{b\left(b + \frac{1}{\Gamma(\beta)}\right)^{1-\beta} - a\left(a + \frac{1}{\Gamma(\beta)}\right)^{1-\beta}}. \tag{2.45}$$

Now, multiplying both sides by

$$\left(c + \frac{1}{\Gamma(\beta)}\right)^{1-\beta}\left[\left(c + \frac{1}{\Gamma(\beta)}\right)^{1-\beta} + \left(c + \frac{1}{\Gamma(\beta)}\right)^{-\beta}\right], \tag{2.46}$$

where

$$\left(c + \frac{1}{\Gamma(\beta)}\right)^{1-\beta} = \frac{f(b) - f(a)}{b\left(b + \frac{1}{\Gamma(\beta)}\right)^{1-\beta} - a\left(a + \frac{1}{\Gamma(\beta)}\right)^{1-\beta}}. \tag{2.47}$$

Now, with f differentiable, we have

$$f'(c) = \lim_{h \to 0} \frac{f(t+h) - f(t)}{h}. \tag{2.48}$$

Now we consider a very small change of variable as follows $h = \left(b + \frac{1}{\Gamma(\beta)}\right)^{1-\beta}$, then we obtain the following expression:

$$^A_0D^\beta_t(f(c)) = \lim_{\varepsilon \to 0} \frac{f\left(c + \varepsilon \left(t_0 + \frac{1}{\Gamma(\beta)}\right)^{1-\beta}\right) - f(c)}{\varepsilon}. \tag{2.49}$$

Then

$$^A_0D^\beta_t(f(c)) = h(\beta, c, a, b)\frac{f(b) - f(a)}{b - a}, \tag{2.50}$$

where

$$h(\beta, c, a, b) = \frac{(b-a)\left[\left(c + \frac{1}{\Gamma(\beta)}\right)^{1-\beta} + \left(c + \frac{1}{\Gamma(\beta)}\right)^{-\beta}\right]}{b\left(b + \frac{1}{\Gamma(\beta)}\right)^{1-\beta} - a\left(a + \frac{1}{\Gamma(\beta)}\right)^{1-\beta}}. \tag{2.51}$$

This completes the proof. $\qquad\qquad\qquad\qquad\qquad\qquad\qquad\square$

Definition 2.3.1. Let $f\colon [a, b] \to \mathbb{R}$ be a continuous function on the closed interval $[a, b]$, then, the 2α-derivative of f is defined as:

$$^A_0D^{2\beta}_t(f(t)) = {}^A_0D^\beta_t\left({}^A_0D^\beta_t(f(t))\right), \quad 0 \le \beta \le 1. \tag{2.52}$$

In general, the $n\beta$-derivative of f is given as:

$$^A_0D^{n\beta}_t(f(t)) = {}^A_0D^\beta_t\left({}^A_0D^{(n-1)\beta}_t(f(t))\right), \quad 0 \le \beta \le 1. \tag{2.53}$$

Remark 1. It is very important to notice that the $n\beta$-derivative of a given function gives information of the previous $n-1$-derivatives of that function. For instance,

$$^A_0D^{2\beta}_t(f(t)) = \left(t + \frac{1}{\Gamma(\beta)}\right)^{1-\beta}\left[(1-\beta)\left(t + \frac{1}{\Gamma(\beta)}\right)^{-\beta}f'\right.$$
$$\left. + \left(t + \frac{1}{\Gamma(\beta)}\right)^{1-\beta}f''\right]. \tag{2.54}$$

This gives this derivative a unique property of memory, which is not provided by any other derivative. It is also easy to verify that if $\beta = 1$, we recover the second derivative of f.

Corollary 2.3.1. *Let $f: [a, b] \to \mathbb{R}$ be a continuous function on the closed interval $[a, b]$. If $\alpha \neq \beta$, then*

$$_0^A D_t^\beta \left(_0^A D_t^\alpha (f(t)) \right) \neq _0^A D_t^\alpha \left(_0^A D_t^\beta (f(t)) \right). \tag{2.55}$$

Proof. In fact:

$$_0^A D_t^\beta \left(_0^A D_t^\alpha (f(t)) \right) = \left(t + \frac{1}{\Gamma(\beta)} \right)^{1-\beta} (1-\alpha) \left(t + \frac{1}{\Gamma(\alpha)} \right)^{-\alpha} f'$$
$$+ \left(t + \frac{1}{\Gamma(\beta)} \right)^{1-\beta} \left(t + \frac{1}{\Gamma(\alpha)} \right)^{1-\alpha} f''.$$

On the other hand, we have

$$_0^A D_t^\alpha \left(_0^A D_t^\beta (f(t)) \right) = \left(t + \frac{1}{\Gamma(\alpha)} \right)^{1-\alpha} (1-\beta) \left(t + \frac{1}{\Gamma(\beta)} \right)^{-\beta} f'$$
$$+ \left(t + \frac{1}{\Gamma(\alpha)} \right)^{1-\alpha} \left(t + \frac{1}{\Gamma(\beta)} \right)^{1-\beta} f''.$$

$\square$

Definition 2.3.2. Let $f: [a, b] \to \mathbb{R}$ be a continuous function on the opened interval (a, b), then, the β-integral of f is given as:

$$_0^A I_t^\beta (f(t)) = \int_0^t \left(x + \frac{1}{\Gamma(\beta)} \right)^{\beta - 1} f(x) \, dx. \tag{2.56}$$

This integral was recently referred to as the Atangana-beta integral.

Theorem 2.3.4. *The fundamental theorem of local β-calculus states that for the first part:*

$$_a^A D_t^\beta \left(_a^A I_t^\beta (f(t)) \right) = f(t), \tag{2.57}$$

with f a given continuous function. For the second part:

$$_a^A I_t^\beta \left(_a^A D_t^\beta (f(t)) \right) = f(t) - f(a). \tag{2.58}$$

for all $x \geq a$ with f a given differentiable function.

Proof. We shall start with part 1. Let f be a continuous function on (a, b), and let $F(t) = {}^{A}_{0}I^{\beta}_{t}(f(t))$, then by definition,

$${}^{A}_{a}D^{\beta}_{t}\left({}^{A}_{a}I^{\beta}_{t}(f(t))\right)$$

$$= \lim_{\varepsilon \to 0} \frac{F\left(t + \varepsilon \left(t + \frac{1}{\Gamma(\beta)}\right)^{1-\beta}\right) - F(t)}{\varepsilon}$$

$$= \lim_{\varepsilon \to 0} \frac{\int_{a}^{\left(t+\varepsilon\left(t+\frac{1}{\Gamma(\beta)}\right)^{1-\beta}\right)} \left(x + \frac{1}{\Gamma(\beta)}\right)^{\beta-1} f(x)\, dx - \int_{a}^{t} f(x) \left(x + \frac{1}{\Gamma(\beta)}\right)^{\beta-1}\, dx}{\varepsilon}$$

$$= \lim_{h \to 0} \frac{\int_{a}^{(t+h)} f(x)\, dx - \int_{a}^{t} f(x)\, dx}{h}$$

$$= f(t). \tag{2.59}$$

$\square$

Second part:

Proof. We shall start with part 1. Let f be a differentiable function on (a, b), and let $h(t) = {}^{A}_{0}D^{\beta}_{t}(f(t))$, then by definition,

$${}^{A}_{a}I^{\beta}_{t}\left({}^{A}_{a}D^{\beta}_{t}(f(t))\right)$$

$$= \int_{a}^{t} \left(x + \frac{1}{\Gamma(\beta)}\right)^{\beta-1} h(x)\, dx$$

$$= \int_{a}^{t} \left(x + \frac{1}{\Gamma(\beta)}\right)^{\beta-1} \left(\lim_{\varepsilon \to 0} \frac{f\left(x + \varepsilon \left(x + \frac{1}{\Gamma(\beta)}\right)^{1-\beta}\right) - f(x)}{\varepsilon}\right) dx$$

$$= \int_{a}^{t} \left(\lim_{h \to 0} \frac{f(x+h) - f(x)}{h}\right) dx$$

$$= f(t) - f(a). \tag{2.60}$$

This completes the proof.

$\square$

Theorem 2.3.5. *Let f be a continuous real function and β-integrable on an opened interval (a, b), then, we can find a real number c such that:*

$${}^{A}_{a}I^{\beta}_{t}(f(t)) = f(c)(b - a)h(a, b). \tag{2.61}$$

Proof. By the extreme value theorem, we can find two real numbers N and $M \in (a, b)$ such that:

$$f(N) = \min_{t \in (a,b)} f(t) \leq f(t) \leq f(M) = \max_{t \in (a,b)} f(t). \tag{2.62}$$

Applying the β-integral on the above inequality Equation (2.79), we obtain the following:

$$\int_a^b \left(t + \frac{1}{\Gamma(\beta)} \right)^{\beta-1} f(N) \, dt \leq \int_a^b \left(t + \frac{1}{\Gamma(\beta)} \right)^{\beta-1} f(t) \, dt$$
$$\leq \int_a^b \left(t + \frac{1}{\Gamma(\beta)} \right)^{\beta-1} f(M) \, dt. \tag{2.63}$$

After integration of the left-hand and right-hand sides of inequality Equation (2.63), we obtain

$$f(N)\frac{\left(b + \frac{1}{\Gamma(\beta)} \right)^{\beta} - \left(a + \frac{1}{\Gamma(\beta)} \right)^{\beta}}{\beta} \leq \int_a^b \left(t + \frac{1}{\Gamma(\beta)} \right)^{\beta-1} f(t) \, dt$$
$$\leq \frac{\left(b + \frac{1}{\Gamma(\beta)} \right)^{\beta} - \left(a + \frac{1}{\Gamma(\beta)} \right)^{\beta}}{\beta} F(M). \tag{2.64}$$

For ease, let

$$I(a, b) = \frac{\left(b + \frac{1}{\Gamma(\beta)} \right)^{\beta} - \left(a + \frac{1}{\Gamma(\beta)} \right)^{\beta}}{\beta}. \tag{2.65}$$

Dividing Equation (2.66) by $I(a, b)$, we obtain

$$f(N) \leq \frac{1}{I(a, b)} \int_a^b \left(t + \frac{1}{\Gamma(\beta)} \right)^{\beta-1} f(t) \, dt \leq F(M). \tag{2.66}$$

However, making use of the intermediate value theorem, we can find a real number $c \in (a, b)$ such that,

$$\frac{1}{I(a, b)} \int_a^b \left(t + \frac{1}{\Gamma(\beta)} \right)^{\beta-1} f(t) \, dt = f(c). \tag{2.67}$$

After manipulations, we obtain

$$_a^A I_t^{\beta}(f(t)) = f(c)(b - a)h(a, b), \quad h(a, b) = \frac{1}{I(a, b)(b - a)}. \tag{2.68}$$

This completes the proof.   □

2.4 DEFINITION OF PARTIAL DERIVATIVE WITH NEW PARAMETER

In this section, we present some useful definition of partial β-derivatives.

Definition 2.4.1. Let f be a function of two variables x and y, then, the β-derivative of f respect to x is defined as follows:

$$
{}_0^A D_x^\beta \left(f(x,y) \right) = \lim_{\varepsilon \to 0} \frac{f\left(x + \varepsilon \left(x + \frac{1}{\Gamma(\beta)} \right)^{1-\beta}, y \right) - f(x,y)}{\varepsilon}. \tag{2.69}
$$

Definition 2.4.2. Let x, y be a system of Cartesian coordinates in two-dimensional Euclidean space, and let i, j be the corresponding basis of unit vectors. The β-divergence of a continuously differentiable vector field $F = Ui + Vj$ is equal to the scalar-valued function:

$$
\text{div}^\beta \ \mathbf{F} = {}_0^A \nabla^\beta \cdot \mathbf{F} = {}_0^A D_x^\beta (U) + {}_0^A D_y^\beta (U). \tag{2.70}
$$

Although expressed in terms of coordinates, the result is invariant under orthogonal transformations, as the physical interpretation suggests. The mixed (β, α)-divergence of F is defined as:

$$
\text{div}^{\beta,\alpha} \ \mathbf{F} = {}_0^A \nabla^{\beta,\alpha} \cdot \mathbf{F} = {}_0^A D_x^\beta (U) + {}_0^A D_y^\alpha (U). \tag{2.71}
$$

Definition 2.4.3. Let x, y be a system of Cartesian coordinates in two-dimensional Euclidean space, and let i, j be the corresponding basis of unit vectors. The β-gradient of a continuously differentiable function f is equal to the vector field:

$$
\text{grad}^\beta f = {}_0^A \nabla^\beta f = {}_0^A D_x^\beta (f(x,t))i + {}_0^A D_y^\beta (f(x,y))j. \tag{2.72}
$$

The mixed (β, α)-grad of f is defined as:

$$
\text{grad}^{\beta,\alpha} f = {}_0^A \nabla^{\beta,\alpha} f = {}_0^A D_x^\beta (f)i + {}_0^A D_y^\alpha (f)j. \tag{2.73}
$$

Definition 2.4.4. The β-Laplace operator in two dimensions of a function f is given by

$$
{}_0^A \Delta^\beta f = \frac{\partial^{2\beta} f(x,y)}{\partial x^{2\beta}} + \frac{\partial^{2\beta} f(x,y)}{\partial y^{2\beta}}, \tag{2.74}
$$

where x and y are the standard Cartesian coordinates of the xy-plane. The mixed (β, α)-Laplace transform method is defined as:

$$\substack{A\\0}\Delta^\beta f = \frac{\partial^{2\beta} f(x,y)}{\partial x^{2\beta}} + \frac{\partial^{2\alpha} f(x,y)}{\partial y^{2\alpha}}. \tag{2.75}$$

Definition 2.4.5. In Cartesian coordinates, the β-curl of a continuously vector field F is, for F, composed of $[Fx, Fy, Fz]$:

$$\begin{vmatrix} \mathbf{i} & \mathbf{j} & \mathbf{k} \\ \frac{\partial^\beta}{\partial x^\beta} & \frac{\partial^\beta}{\partial y^\beta} & \frac{\partial^\beta}{\partial z^\beta} \\ F_x & F_y & F_z \end{vmatrix},$$

where $i, j,$ and k are the unit vectors for the x-, y-, and z-axes, respectively. This expands as follows:

$$\left(\frac{\partial^\beta F_z}{\partial y^\beta} - \frac{\partial^\beta F_y}{\partial z^\beta}\right)\mathbf{i} + \left(\frac{\partial^\beta F_x}{\partial^\beta z} - \frac{\partial^\beta F_z}{\partial^\beta x}\right)\mathbf{j} + \left(\frac{\partial^\beta F_y}{\partial x^\beta} - \frac{\partial^\beta F_x}{\partial y^\beta}\right)\mathbf{k}. \tag{2.76}$$

Although expressed in terms of coordinates, the result is invariant under proper rotations of the coordinate axes but the result inverts under reflection.

2.5 PROPERTIES OF PARTIAL BETA-DERIVATIVES

In this section, we present some useful properties and theorems in connection of the beta-partial derivative. We shall start with some fundamental theorems of partial differentiation.

Theorem 2.5.1. *Clairaut's theorem for partial beta-derivatives: Assume that $f(x,y)$ is function which $\partial_x^\beta\left[\partial_y^\alpha(f(x,y))\right]$ and $\partial_y^\alpha\left[\partial_x^\beta(f(x,y))\right]$ exist and are continues over the domain $D \subset \mathbb{R}_2$ then,*

$$\partial_x^\beta\left[\partial_y^\alpha(f(x,y))\right] = \partial_y^\alpha\left[\partial_x^\beta(f(x,y))\right]. \tag{2.77}$$

Proof. By definition as in equation, we have

$$\partial_x^\beta\left[\partial_y^\alpha(f(x,y))\right] = \partial_x^\beta\left[\lim_{\varepsilon\to0}\frac{f\left(x,y+\varepsilon\left(y+\frac{1}{\Gamma(\beta)}\right)^{1-\alpha}\right) - f(x,y)}{\varepsilon}\right]. \tag{2.78}$$

Letting $\varepsilon \left(y + \frac{1}{\Gamma(\beta)}\right)^{1-\alpha} = k$, we obtain

$$\partial_x^{\beta}\left[\partial_y^{\alpha}(f(x,y))\right] = \left(y + \frac{1}{\Gamma(\alpha)}\right)^{1-\alpha} \partial_x^{\beta}\left[\lim_{k\to 0} \frac{f(x,y+k) - f(x,y)}{k}\right].$$
(2.79)

Since f is differentiable in y-direction, Equation (2.79) becomes:

$$\partial_x^{\beta}\left[\partial_y^{\alpha}(f(x,y))\right] = \left(y + \frac{1}{\Gamma(\alpha)}\right)^{1-\alpha} \partial_x^{\beta}\left[\frac{\partial f(x,y)}{\partial y}\right].$$
(2.80)

However, using again the definition of beta-derivative, we transform Equation (2.80) to:

$$\partial_x^{\beta}\left[\partial_y^{\alpha}(f(x,y))\right] = \left(y + \frac{1}{\Gamma(\alpha)}\right)^{1-\alpha}$$
$$\left[\lim_{\varepsilon\to 0} \frac{\frac{\partial f}{\partial y}\left(x + \varepsilon\left(x + \frac{1}{\Gamma(\beta)}\right)^{1-\beta}, y\right) - \frac{f}{\partial y}(x,y)}{\varepsilon}\right].$$
(2.81)

Again letting $\varepsilon \left(x + \frac{1}{\Gamma(\alpha)}\right)^{1-\beta} = l$, we obtain

$$\partial_x^{\beta}\left[\partial_y^{\alpha}(f(x,y))\right] = \left(y + \frac{1}{\Gamma(\alpha)}\right)^{1-\alpha}\left(x + \frac{1}{\Gamma(\beta)}\right)^{1-\beta}$$
$$\left[\lim_{l\to 0} \frac{\frac{\partial f}{\partial y}(x+l,y) - \frac{f}{\partial y}(x,y)}{l}\right].$$
(2.82)

Since f is differentiable in y-direction, Equation (2.82) becomes:

$$\partial_x^{\beta}\left[\partial_y^{\alpha}(f(x,y))\right] = \left(y + \frac{1}{\Gamma(\alpha)}\right)^{1-\alpha}\left(x + \frac{1}{\Gamma(\beta)}\right)^{1-\beta}\left[\frac{\partial^2 f(x,y)}{\partial x \partial y}\right]. \quad (2.83)$$

Since f is continuous, we use Clairaut's theorem for partial derivatives, which is

$$\frac{\partial^2 f(x,y)}{\partial x \partial y} = \frac{\partial^2 f(x,y)}{\partial y \partial x}.$$
(2.84)

Nevertheless, replacing Equation (2.84) into Equation (2.83), we obtain

$$\partial_x^\beta \left[\partial_y^\alpha (f(x,y)) \right] = \left(y + \frac{1}{\Gamma(\alpha)} \right)^{1-\alpha} \left(x + \frac{1}{\Gamma(\beta)} \right)^{1-\beta} \left[\frac{\partial^2 f(x,y)}{\partial y \partial x} \right]. \quad (2.85)$$

Using the definition of the first principle for derivative, Equation (2.85) can be reformulated as:

$$\partial_x^\beta \left[\partial_y^\alpha (f(x,y)) \right] = \left(y + \frac{1}{\Gamma(\alpha)} \right)^{1-\alpha} \left(x + \frac{1}{\Gamma(\beta)} \right)^{1-\beta}$$
$$\left[\lim_{l \to 0} \frac{\frac{\partial f}{\partial x}(x+l,y) - \frac{f}{\partial x}(x,y)}{l} \right]. \quad (2.86)$$

Again letting $\varepsilon \left(x + \frac{1}{\Gamma(\alpha)} \right)^{1-\beta} = l$, we obtain

$$\partial_x^\beta \left[\partial_y^\alpha (f(x,y)) \right] = \left(x + \frac{1}{\Gamma(\beta)} \right)^{1-\beta}$$
$$\left[\lim_{\varepsilon \to 0} \frac{\frac{\partial f}{\partial x}\left(x + \varepsilon \left(x + \frac{1}{\Gamma(\alpha)} \right)^{1-\alpha}, y \right) - \frac{f}{\partial x}(x,y)}{\varepsilon} \right]. \quad (2.87)$$

Using the definition of beta-derivative, we obtain

$$\partial_x^\beta \left[\partial_y^\alpha (f(x,y)) \right] = \left(x + \frac{1}{\Gamma(\beta)} \right)^{1-\beta} \partial_y^\alpha \left[\frac{\partial f(x,y)}{\partial x} \right]. \quad (2.88)$$

Using the first principle, Equation (2.88) becomes:

$$\partial_x^\beta \left[\partial_y^\alpha (f(x,y)) \right] = \left(x + \frac{1}{\Gamma(\beta)} \right)^{1-\beta} \partial_y^\alpha \left[\lim_{l \to 0} \frac{\frac{\partial f}{\partial x}(x+k,y) - \frac{f}{\partial x}(x,y)}{k} \right]. \quad (2.89)$$

Letting $\varepsilon \left(y + \frac{1}{\Gamma(\beta)} \right)^{1-\alpha} = k$, we obtain

$$\partial_x^\beta \left[\partial_y^\alpha (f(x,y)) \right] = \partial_y^\alpha \left[\lim_{\varepsilon \to 0} \frac{f\left(x + \varepsilon \left(x + \frac{1}{\Gamma(\beta)} \right)^{1-\alpha}, y \right) - f(x,y)}{\varepsilon} \right]. \quad (2.90)$$

Thus we obtain the requested result:

$$\partial_x^\beta \left[\partial_y^\alpha (f(x,y)) \right] = \partial_y^\alpha \left[\partial_x^\beta (f(x,y)) \right]. \quad (2.91)$$

This completes the proof.   □

The beta-derivative satisfies the following properties:

1.
$$_0^A\Delta^\beta \cdot (af + bg) = a_0^A\Delta^\beta \cdot (f) + b_0^A\Delta^\beta (g). \tag{2.92}$$

2.
$$_0^A\Delta^\beta \cdot (aF + bG) = a_0^A\Delta^\beta \cdot (F) + b_0^A\Delta^\beta \cdot (G). \tag{2.93}$$

3.
$$_0^A\Delta^\beta \times (aF + bG) = a_0^A\Delta^\beta (F) + b_0^A\Delta^\beta \times (G). \tag{2.94}$$

4.
$$_0^A\Delta^\beta \cdot (_0^A\Delta^\beta \times F) = 0. \tag{2.95}$$

5.
$$_0^A\Delta^\beta \times (_0^A\Delta^\beta \cdot F) = 0. \tag{2.96}$$

The first three properties are a direct effect of the linear representative of the different operations and are fairly trivial to launch. However, the last two relations rely on Clairaut's result established in theorem. We shall present the proof of relation 4, from the definition of the beta-divergence and beta-curl.

Proof. In fact, property 4 can be expressed as follows:

$$_0^A\Delta^\beta \cdot (_0^A\Delta^\beta \times F) = \sum_{i=1}^{3}\sum_{j=1}^{3}\sum_{k=1}^{3} e_{ijk}\partial_{x_i}^\beta \Delta_{x_j}^\beta F_k, \tag{2.97}$$

where e_{ijk} is the Levi-Civita symbol and is equal to zero whenever two or more indexes take the same value. The triple summation in Equation (2.97) reduces to

$$\sum_{i=1}^{3}\sum_{j=1}^{3}\sum_{k=1}^{3} e_{ijk}\partial_{x_i}^\beta \Delta_{x_j}^\beta F_k = \sum_{i=1}^{\beta} \left(e_{ijk}\partial_{x_i}^\beta \left[\partial_{x_j}^\beta F_k - \partial_{x_k}^\beta F_j\right]\right), \quad i \neq j \neq k. \tag{2.98}$$

Or the above can be converted to

$$_0^A\Delta^\beta \cdot (_0^A\Delta^\beta \times F) = \partial_{x_i}^\beta \left[\partial_{x_j}^\beta F_k - \partial_{x_k}^\beta F_j\right] + \partial_{x_j}^\beta \left[\partial_{x_k}^\beta F_i - \partial_{x_i}^\beta F_k\right]$$
$$+ \partial_{x_k}^\beta \left[\partial_{x_i}^\beta F_j - \partial_{x_j}^\beta F_i\right] = 0, \tag{2.99}$$

due to a direct application of the beta-Clairaut's theorem. The last property is established using similar arguments as this one.   □

Vector calculus or vector analysis is a branch of mathematics concerned with differentiation and integration of vector fields, primarily in three-dimensional Euclidean space $\mathbb{R}^3$. The term "vector calculus" is sometimes used as a synonym for the broader subject of multivariable calculus, which includes vector calculus as well as partial differentiation and multiple integration. Vector calculus plays an important role in differential geometry and in the study of partial differential equations. It is used extensively in physics and engineering, especially in the description of electromagnetic fields, gravitational fields, and fluid flow. The basic objectives are, but are not limited to, the follow:

1. *Scalar fields.* A scalar field associates a scalar value to every point in a space. The scalar may either be a mathematical number or a physical quantity. Examples of scalar fields in applications include the temperature distribution throughout space, the pressure distribution in a fluid, and spin-zero quantum fields, such as the Higgs field. These fields are the subject of scalar field theory.
2. *Vector fields.* A vector field is an assignment of a vector to each point in a subset of space. A vector field in the plane, for instance, can be visualized as a collection of arrows with a given magnitude and direction each attached to a point in the plane. Vector fields are often used to model, for example, the speed and direction of a moving fluid throughout space, or the strength and direction of some force, such as the magnetic or gravitational force, as it changes from point to point.

Vector calculus was developed from quaternion analysis by J. Willard Gibbs and Oliver Heaviside near the end of the nineteenth century, and most of the notation and terminology was established by Gibbs and Edwin Bidwell Wilson in their 1901 book, *Vector Analysis*. In the conventional form, using cross products, vector calculus does not generalize to higher dimensions, while the alternative approach of geometric algebra, which uses exterior products, does generalize, as discussed below.

Theorem 2.5.2. *For remembrance, divergence theorem states that, let the vector field $\underline{F}$ have a continuous derivative on an open region space D containing the volume V and surface S of V positively outward orientated, then,*

$$\iiint \Delta \cdot \underline{F} \, dV = \iint \underline{F} \cdot \underline{n} \, dS. \tag{2.100}$$

Therefore within the framework of the beta-derivatives, this theorem can be rephrased; we shall first present the following definition.

Definition 2.5.1. Let the vector field $\underline{F}$ having the partial β-derivatives with respect to all variable $\underline{x} = (x_1, \ldots, x_m)^T$ on D. Then we denote by $H_{\underline{x}}^{\beta}\underline{F}$ the vector defined as:

$$H_{\underline{x}}^{\beta}\underline{F} = \sum_{i=1}^{m} \left[e_{x_i}^T \left[\Delta_{\underline{x}}^{\beta}(\underline{F})^T \right] \cdot \underline{e}_{x_i} \right] \underline{e}_{x_i} = \sum_{i=1}^{m} \partial_{x_i}^{\beta} F_{x_i} \underline{e}_{x_i}. \qquad (2.101)$$

With this Definition 2.5.1, we propose the following theorem.

Theorem 2.5.3. *Let the vector field $\underline{F}$ have the continuous partial beta-derivatives on an open region of the space D containing the volume V and S is the boundary surface of V positively outward oriented. Then we have the following relation:*

$$\iiint \Delta_{\underline{x}}^{\beta} \cdot \underline{F} \, dV^{\beta} = \iint_S H_{\underline{x}}^{\beta-1} \underline{F} \cdot \underline{n} \, dS^{\beta}, \qquad (2.102)$$

with

$$dV^{\beta} = \left(x + \frac{1}{\Gamma(\beta)} \right)^{\beta-1} \left(y + \frac{1}{\Gamma(\beta)} \right)^{\beta-1} \left(z + \frac{1}{\Gamma(\beta)} \right)^{\beta-1} dx \, dy \, dz \quad and$$

$$dS^{\beta} = \left(x + \frac{1}{\Gamma(\beta)} \right)^{\beta-1} \left(y + \frac{1}{\Gamma(\beta)} \right)^{\beta-1} dx \, dy. \qquad (2.103)$$

Proof. To prove this theorem, we first notice that

$$\Delta_{\underline{x}}^{\beta} \cdot \underline{F} = \Delta_{\underline{x}} \cdot H_{\underline{x}}^{\beta-1} \underline{F}, \qquad (2.104)$$

which was established in Definition 2.5.1. We then have, by direct application of the classical version of the divergence theorem,

$$\iiint_V \Delta_{\underline{x}}^{\beta} \cdot \underline{F} \, dV^{\beta} = \iint_S H_{\underline{x}}^{\beta-1} \underline{F} \cdot \underline{n} \, dS^{\beta}. \qquad (2.105)$$

This completes the proof. $\qquad\qquad\qquad\qquad\qquad\qquad\qquad\qquad\Box$

In an analogous way, the distinguished theorem of Green and Stokes of the vector calculus can perhaps be appropriately modified to acclimatize straightforwardly the perception of beta-derivative; we therefore have the following theorems.

Theorem 2.5.4. *Green's theorem for β-derivative. Let C be a simple positively oriented, piecewise smooth and closed curve in* $\mathbb{R}_2$, *says for instance the* $x - y$-*plane, furthermore assume D is the interior of C. If* $f(x,y)$ *and* $g(x,y)$ *are two functions having continuous partial conformable fractional derivative on D then,*

$$\iint_D (\partial_x^\beta g - \partial_y^\beta f) = \int_C \partial_y^{\beta-1} f\, dx^\beta + \partial_x^{\beta-1}\, dy^\beta. \tag{2.106}$$

Proof. This is clearly a direct application of the classical version of Green's theorem. Now since

$$\iint_D (\partial_x^\beta g - \partial_y^\beta f) = \iint_D \partial_x(\partial_x^{\beta-1} g) - \partial_y(\partial_y^{\beta-1})\, dS^\beta, \tag{2.107}$$

applying the Green function theorem yields:

$$\iint_D (\partial_x^\beta g - \partial_y^\beta f) = \int_C \partial_y^{\beta-1} f\, dx^\beta + \partial_x^{\beta-1}\, dy^\beta. \tag{2.108}$$

□

Remember for supplementary determinations, this relation into the plane can be articulated in terms of the conformable fractional of the vector field $\underline{F} = (f, g, h)$, h being an arbitrary function in D.

Theorem 2.5.5. *Stokes's theorem for β-derivative. Let S be a regular surface of class* $\mathbb{C}^2$ *described by the parametric equations* $P(u,v) = (x(u,v),$ $y(u,v), z(u,v))$, $u, v \in S$, *where x, y, and z are the Cartesian coordinates. Consider a simple positively oriented, piecewise smooth, closed curve* C_k *in the plane* $u - v$, *assume K to be the interior of* C_k. *S and C are images of the domain K and its boundary* C_k *in the Cartesian space and* $\underline{F}$ *is a vector field having continuous partial β-derivatives on S then*

$$\iint_S \left[\Delta_{\underline{x}}^\beta \times \underline{F} + \left(\partial_x\left[K_{\underline{x}}^{\beta-1}\right]\cdot F_y e_y\right) e_x + \left(\partial_y\left[K_{\underline{y}}^{\beta-1}\right]\cdot F_z e_z\right) e_y\right.$$
$$\left. + \left(\partial_z\left[K_{\underline{z}}^{\beta-1}\right]\cdot F_x e_x\right) e_z\right] = \int_C \partial_z^{\beta-1} F_x\, dx^\beta + \partial_x^{\beta-1}\, dy^\beta + \partial_y^{\beta-1} F_z\, dz^{\beta-1}, \tag{2.109}$$

with

$$K_{\underline{x}}^{\beta-1} = -\Delta_{\underline{x}}^\beta = (\partial_y^\beta - \partial_z^\beta) e_x + (\partial_z^\beta - \partial_x^\beta) e_y + (\partial_x^\beta - \partial_y^\beta) e_z, \tag{2.110}$$

where $\underline{n}$ *is the unit positive outside normal to S.*

CHAPTER 3

Novel integrals transform

In mathematics, an integral transform is any transform T of the following form:

$$(Tf)(u) = \int_a^b K(t, u)f(t) \, dt. \tag{3.1}$$

The input of this transform is a function f, and the output is another function Tf. An integral transform is a particular kind of mathematical operator. There are numerous useful integral transforms. Each is specified by a choice of the function K of two variables, the kernel function or nucleus of the transform. Here are some transform operator that one can find in the literature: Buschman transform, Fourier transform, Fourier–Stieltjes transform, G-transform, H-transform, Hadamard transform, Hankel transform, Hartley transform, Hough transform, Kontorovich–Lebedev transform, Mehler–Fock transform, Meijer transform, Mellin transform, Narain G-transform, Operational Mathematics, Radon transform, Stieltjes transform, W-transform, Wavelet transform, and Z-transform. We can find the formulation of these integral transforms in [61–69]. We shall present the definition of some useful integral transform operators.

3.1 DEFINITION OF SOME INTEGRAL TRANSFORM OPERATORS

Definition 3.1.1. The Laplace transform is a widely used integral transform in mathematics and electrical engineering named after Pierre–Simon Laplace that transforms a function of time into a function of complex frequency. The Laplace transform is named after mathematician and astronomer Pierre–Simon Laplace, who used a similar transform (now called Z-transform) in his work on probability theory. The current widespread use of the transform came about soon after World War II although it had been used in the nineteenth century by Abel, Lerch, Heaviside, and Bromwich. The Laplace transform of a function $f(t)$, defined for all real numbers $0 \leq t$, is the function $F(s)$, defined by:

$$F(s) = \int_0^\infty f(t) \, e^{-st} \, dt. \tag{3.2}$$

The parameter s is the complex number frequency $s = \sigma + iw$.

Derivative with a New Parameter. http://dx.doi.org/10.1016/B978-0-08-100644-3.00003-9
© 2016 Elsevier Ltd. All rights reserved.

Definition 3.1.2. The Fourier transform decomposes a function of time (a signal) into the frequencies that make it up, similarly to how a musical chord can be expressed as the amplitude (or loudness) of its constituent notes. The Fourier transform of a function of time itself is a complex-valued function of frequency, whose absolute value represents the amount of that frequency present in the original function, and whose complex argument is the phase offset of the basic sinusoid in that frequency. The Fourier transform of a function $f(t)$, defined for all real numbers $0 \leq t$, is the function $F(s)$, defined by:

$$F(s) = \int_{-\infty}^{\infty} e^{-2\pi i s t} f(t) \, dt. \tag{3.3}$$

The parameter s is the complex number frequency $s = \sigma + iw$.

Definition 3.1.3. The Sumudu transform is an integral transform similar to the Laplace transform, introduced in the early 1990s by Watugala [70] to solve differential equations and control engineering problems. The Sumudu transform of a function f, defined for all real numbers t 0, is the function $F_s(u)$, defined by:

$$S(f(t)) = F_s(u) = \int_0^{\infty} \frac{1}{u} \exp\left[-\frac{t}{u}\right] f(t) \, dt. \tag{3.4}$$

Definition 3.1.4. The Mellin transform is an integral transform that may be regarded as the multiplicative version of the two-sided Laplace transform. This integral transform is closely connected to the theory of Dirichlet series, and is often used in number theory, mathematical statistics, and the theory of asymptotic expansions; it is closely related to the Laplace transform and the Fourier transform, and the theory of the gamma function and allied special functions. The Mellin transform of a function f is:

$$\{\mathcal{M}f\}(s) = \varphi(s) = \int_0^{\infty} x^{s-1} f(x) \, dx. \tag{3.5}$$

The inverse transform is:

$$\left\{\mathcal{M}^{-1}\varphi\right\}(x) = f(x) = \frac{1}{2\pi i} \int_{c-i\infty}^{c+i\infty} x^{-s} \varphi(s) \, ds. \tag{3.6}$$

The notation implies this is a line integral taken over a vertical line in the complex plane. Conditions under which this inversion is valid are given in the Mellin inversion theorem. The transform is named after the Finnish mathematician Hjalmar Mellin.

Definition 3.1.5. The Radon transform in two dimensions, named after the Austrian mathematician Johann Radon, is the integral transform consisting of the integral of a function over straight lines. The transform was introduced in 1917 by Radon [71], who also provided a formula for the inverse transform. Let $(x) = (x,y)$ be a compactly supported continuous function on $\mathbb{R}^2$. The Radon transform, R, is a function defined on the space of straight lines L in $\mathbb{R}^2$ by the line integral along each such line:

$$Rf(L) = \int_L f(\mathbf{x}) \,|d\mathbf{x}|. \tag{3.7}$$

Definition 3.1.6. Nowadays, wavelet transformation is one of the most popular candidates of the time-frequency transformations. The integral wavelet transform is the integral transform defined as:

$$\left[W_\psi f\right](a,b) = \frac{1}{\sqrt{|a|}} \int_{-\infty}^{\infty} \overline{\psi\left(\frac{x-b}{a}\right)} f(x)\,dx. \tag{3.8}$$

The wavelet coefficients c_{jk} are then given by:

$$c_{jk} = \left[W_\psi f\right]\left(2^{-j}, k2^{-j}\right). \tag{3.9}$$

Here, $a = 2^{-j}$ is called the binary dilation or dyadic dilation, and $b = k2^{-j}$ is the binary or dyadic position.

Definition 3.1.7. The Hankel transform expresses any given function $f(r)$ as the weighted sum of an infinite number of Bessel functions of the first kind $J(kr)$. More precisely, the Hankel transform of order of a function $f(r)$ is given by:

$$F_\nu(k) = \int_0^{\infty} f(r)J_\nu(kr)\, r\,dr, \tag{3.10}$$

where J is the Bessel function of the first kind of order with $\geq \frac{1}{2}$. The inverse Hankel transform of $F(k)$ is defined as:

$$f(r) = \int_0^{\infty} F_\nu(k)J_\nu(kr)k\,dk. \tag{3.11}$$

Definition 3.1.8. The Weierstrass transform [72] of a function $f\colon \mathbb{R} \to \mathbb{R}$, named after Karl Weierstrass, is the function F defined by:

$$F(x) = \frac{1}{\sqrt{4\pi}} \int_{-\infty}^{\infty} f(y)\ e^{-\frac{(x-y)^2}{4}}\ dy = \frac{1}{\sqrt{4\pi}} \int_{-\infty}^{\infty} f(x-y)\ e^{-\frac{y^2}{4}}\ dy, \tag{3.12}$$

the convolution of f with the Gaussian function $\frac{1}{\sqrt{4\pi}}\,e^{-x^2/4}$. Instead of $F(x)$ we also write $W[f](x)$. Note that $F(x)$ need not exist for every real number x, because the defining integral may fail to converge.

Definition 3.1.9. The N-transform is an integral transform similar to the Laplace transform and Sumudu transform, introduced by Zafar Hayat Khan [73] in 2008. It converges to both the Laplace and Sumudu transforms just by changing variables. The natural transform of a function $f(t)$, defined for all real numbers $t \geq 0$, is the function $R(u, s)$, defined by:

$$R(u, s) = \mathcal{N}\{f(t)\} = \int_0^{\infty} f(ut)\, e^{-st}\, dt. \qquad (3.13)$$

Khan showed that the above integral converges to Laplace transform when $u = 1$, and into Sumudu transform for $s = 1$.

Definition 3.1.10. The Hartley transform is an integral transform closely related to the Fourier transform, but which transforms real-valued functions to real-valued functions. The Hartley transform of a function $f(t)$ is defined by:

$$H(\omega) = \{\mathcal{H}f\}\,(\omega) = \frac{1}{\sqrt{2\pi}} \int_{-\infty}^{\infty} f(t)\, \mathrm{cas}(\omega t)\, dt, \qquad (3.14)$$

where ω can in applications be an angular frequency and

$$\mathrm{cas}(t) = \cos(t) + \sin(t) = \sqrt{2}\,\sin(t + \pi/4) = \sqrt{2}\,\cos(t - \pi/4) \quad (3.15)$$

is the cosine-and-sine or Hartley kernel.

The above integral transforms were proposed within the scope of calculus. We shall present in the next sections some integral transform associate to beta-calculus.

3.2 DEFINITION AND PROPERTIES OF THE BETA-LAPLACE TRANSFORM

Definition 3.2.1. Let g be a function defined in $(0, \infty)$. We then define the beta-Laplace transform of g as:

$$\mathcal{L}(g(x))(s) = \int_0^{\infty} \left(x + \frac{1}{\Gamma(1 - \beta)}\right)^{\beta - 1} e^{-sx} f(x)\, dx. \qquad (3.16)$$

We shall give some properties of the above operator.

3.2.1 Properties of the beta-Laplace transform

1. Linearity:

$$\mathcal{L}(af(x) + bg(x))(s) = a\mathcal{L}(f) + b\mathcal{L}(g). \tag{3.17}$$

2. Delay:

$$\mathcal{L}({}_{0}^{A}D_{t}^{\beta}(f(x-a)\delta(t-a)))(s) = \mathcal{L}(f)s\,e^{-as}. \tag{3.18}$$

3. Beta-derivative of first derivative:

$$\mathcal{L}({}_{0}^{A}D_{t}^{\beta}(\partial_{t}f(t)))(s) = s^{2}\mathcal{L}(f(t))(s) - sf(0) - f(0). \tag{3.19}$$

4. Beta-derivative of nth derivative:

$$\mathcal{L}({}_{0}^{A}D_{t}^{\beta}(\partial_{t}^{n}f(t)))(s) = s^{n+1}\mathcal{L}(f(t))(s) - \sum_{j=0}^{n} s^{j}f^{(n-1)}(0). \tag{3.20}$$

5. Beta-derivative of the Caputo derivative:

$$\mathcal{L}({}_{0}^{A}D_{t}^{\beta}({}_{0}^{C}D_{t}^{\alpha}f(t)))(s) = s^{\alpha+1}\mathcal{L}(f(t))(s) - \sum_{j=0}^{n} s^{\alpha-j}f^{(n-1)}(0), \quad n-1 < \alpha < n. \tag{3.21}$$

6. Integral:

$$\mathcal{L}\left({}_{0}^{A}D_{t}^{\beta}\left(\int_{0}^{t} f(x)\,dx\right)(s)\right) = \mathcal{L}(f(t)). \tag{3.22}$$

7. Convolution:

$$\mathcal{L}({}_{0}^{A}D_{t}^{\beta}(f * g)) = s\mathcal{L}(f(t))\mathcal{L}(g(t)). \tag{3.23}$$

8. Complex shift:

$$\mathcal{L}({}_{0}^{A}D_{t}^{\beta}(f(t)\,e^{-at})) = s\mathcal{L}(f(t))(s+a) - f(0). \tag{3.24}$$

9. Multiplication by distance:

$$\mathcal{L}({}_{0}^{A}D_{t}^{\beta}(f(t)t)) = -s\mathcal{L}(f(t))'(s). \tag{3.25}$$

10. Distance scaling:

$$\mathcal{L}({}_{0}^{A}D_{t}^{\beta}(f(at))) = \frac{s}{a}\mathcal{L}(f(t))\left(\frac{s}{a}\right) - f(0). \tag{3.26}$$

We shall present in detail the proof of the above properties.

Proof. Linearity: By definition, we have

$$
\mathcal{L}(af(x) + bg(x))(s) = \int_0^\infty \left(t + \frac{1}{\Gamma(\beta)}\right)^{\beta-1} e^{-st}(af(t) + bg(t)) \, dt
$$

$$
= a \int_0^\infty \left(t + \frac{1}{\Gamma(\beta)}\right)^{\beta-1} e^{-st} f(t) \, dt
$$

$$
+ b \int_0^\infty \left(t + \frac{1}{\Gamma(\beta)}\right)^{\beta-1} e^{-st} g(t) \, dt
$$

$$
= a\mathcal{L}(f) + b\mathcal{L}(g). \tag{3.27}
$$

This completed the proof. □

Proof. Delay: Using the definition of the variable-order derivative and its anti-derivative, we have

$$
\mathcal{L}({}_0^A D_t^\beta (f(x - a)\delta(t - a)))(s) = \int_0^\infty \left(t + \frac{1}{\Gamma(\beta)}\right)^{\beta-1}
$$

$$
e^{-st}({}_0^A D_t^\beta f(t - a)\delta(t - a)) \, dt, \tag{3.28}
$$

but

$$
{}_0^A D_t^\beta (f(t - a)\delta(t - a))
$$

$$
= \lim_{\varepsilon \to 0} \frac{f\left(t - a + \varepsilon\left(t - a + \frac{1}{\Gamma(\beta)}\right)^{1-\beta}\right) \cdot \delta\left(t - a + \varepsilon\left(t - a + \frac{1}{\Gamma(\beta)}\right)^{1-\beta}\right) - f(t - a)\delta(t - a)}{\varepsilon}.
$$

$$
\tag{3.29}
$$

However, using the product rule for the beta-derivative, we obtain the following:

$$
{}_0^A D_t^\beta (f(t-a)\delta(t-a)) = \delta(t-a){}_0^A D_t^\beta (f(t-a)) + f(t-a){}_0^A D_t^\beta (\delta(t-a)). \tag{3.30}
$$

Thus,

$$
\int_0^\infty \left(t + \frac{1}{\Gamma(\beta)}\right)^{1-\beta} e^{-stA} {}_0 D_t^\beta (f(t - a)\delta(t - a)) \, dt
$$

$$
= \int_0^\infty \frac{d(f(t - a) \cdot \delta(t - a))}{dt} e^{-st} \, dt. \tag{3.31}
$$

Now, using the property of the Laplace transform for the derivative, we have

$$
\int_0^\infty \frac{d(f(t - a) \cdot \delta(t - a))}{dt} e^{-st} \, dt = s \, e^{-as} \mathcal{L}(s), \tag{3.32}
$$

which completes the proof. □

Proof. Using the definition of the variable-order derivative together with the anti-derivative, we obtain the following expression:

$$
\mathcal{L}({}_{0}^{A}D_{t}^{\beta}(\partial_{t}f(t)))(s) = \int_{0}^{\infty} e^{-st}\left(t + \frac{1}{\Gamma(\beta)}\right)^{1-\beta}({}_{0}^{A}D_{t}^{\beta}(\partial_{t}f(t)))\, dt
$$

$$
= \int_{0}^{\infty} e^{-st}\left(t + \frac{1}{\Gamma(\beta)}\right)^{1-\beta}
$$

$$
\left(\lim_{\varepsilon \to 0} \frac{f'\left(t + \varepsilon\left(t + \frac{1}{\Gamma(\beta)}\right)^{1-\beta}\right) - f'(t)}{\varepsilon}\right) dt.
$$

$$(3.33)$$

For simplicity, let $h = \left(t + \frac{1}{\Gamma(\beta)}\right)^{1-\beta}$. The above equation can then be rewritten as:

$$
\mathcal{L}({}_{0}^{A}D_{t}^{\beta}(\partial_{t}f(t)))(s) = \int_{0}^{\infty} e^{-st}\left(t + \frac{1}{\Gamma(\beta)}\right)^{1-\beta}({}_{0}^{A}D_{t}^{\beta}(\partial_{t}f(t)))\, dt
$$

$$
= \int_{0}^{\infty} e^{-st}\left(\lim_{h \to 0} \frac{f'(t+h) - f'(t)}{h}\right) dt
$$

$$
= s^{2}\mathcal{L}(f)(s) - sf(0) - f(0).
$$

$$(3.34)$$

This completes the proof. □

Proof. Using the definition of variable-order derivative together with the anti-derivative, we obtain the following expression

$$
\mathcal{L}({}_{0}^{A}D_{t}^{\beta}(\partial_{t^n}f(t)))(s) = \int_{0}^{\infty} e^{-st}\left(t + \frac{1}{\Gamma(\beta)}\right)^{1-\beta}({}_{0}^{A}D_{t}^{\beta}(\partial_{t^n}f(t)))\, dt
$$

$$
= \int_{0}^{\infty} e^{-st}\left(t + \frac{1}{\Gamma(\beta)}\right)^{1-\beta}
$$

$$
\left(\lim_{\varepsilon \to 0} \frac{f^{n}\left(t + \varepsilon\left(t + \frac{1}{\Gamma(\beta)}\right)^{1-\beta}\right) - f^{n}(t)}{\varepsilon}\right) dt.
$$

$$(3.35)$$

For simplicity, let $h = \left(t + \frac{1}{\Gamma(\beta)}\right)^{1-\beta}$ the above equation can be rewritten as

$$\mathcal{L}(_0^A D_t^\beta (\partial_{t^n}^n f(t)))(s) = \int_0^\infty e^{-st} \left(t + \frac{1}{\Gamma(\beta)}\right)^{1-\beta} (_0^A D_t^\beta (\partial_{t^n}^n f(t)))\, dt$$

$$= \int_0^\infty e^{-st} \left(\lim_{h \to 0} \frac{f^n(t+h) - f^n(t)}{h}\right) dt$$

$$= s^{n+1} \mathcal{L}(f)(s) - \sum_{j=0}^n s^j f^{(n-1)}(0). \tag{3.36}$$

This completes the proof. □

Proof. Integral, using the definition, we have

$$\mathcal{L}_\beta \left(_0^A D_t^\beta \left(\int_0^t f(x)\, dx\right)\right)(s) = \int_0^\infty \left(t + \frac{1}{\Gamma(\beta)}\right)^{\beta-1}$$

$$\times e^{-st} \left(\lim_{\varepsilon \to 0} \frac{\int_0^{t+\varepsilon\left(t+\frac{1}{\Gamma(\beta)}\right)^{1-\beta}} f(t)\, dt - \int_0^t f(t)\, dt}{\varepsilon}\right) dt, \tag{3.37}$$

$h = \left(t + \frac{1}{\Gamma(\beta)}\right)^{1-\beta}$. Equation (3.37) can be rewritten as:

$$\mathcal{L}_\beta \left(_0^A D_t^\beta \left(\int_0^t f(x)\, dx\right)\right)(s) = \int_0^\infty \left(t + \frac{1}{\Gamma(\beta)}\right)^{\beta-1} \left(t + \frac{1}{\Gamma(\beta)}\right)^{1-\beta}$$

$$\times e^{-st} \left(\lim_{h \to 0} \frac{\int_0^{t+h} f(t)\, dt - \int_0^t f(t)\, dt}{h}\right) dt$$

$$= \int_0^\infty e^{-st} \left(\lim_{h \to 0} \frac{\int_0^{t+h} f(t)\, dt - \int_0^t f(t)\, dt}{h}\right) dt$$

$$= \mathcal{L}(f)(s), \tag{3.38}$$

which is the completion of the proof. □

Proof.

$$\mathcal{L}_\beta (_0^A D_t^\beta (f * g(t)))(s) = \int_0^\infty e^{-st} \frac{d(f * g(t))}{dt}\, dt = s\mathcal{L}(f)(s)\mathcal{L}(g)(s). \tag{3.39}$$

□

3.3 DEFINITION AND PROPERTIES OF THE BETA-SUMUDU TRANSFORM

We introduce in this section a connected Sumudu transform of beta-calculus.

Definition 3.3.1. Let f be a function defined in $(0, \infty)$, then, we defined the beta-Sumudu transform of f as:

$$S_\beta(f(x))(u) = \int_0^\infty \left(t + \frac{1}{\Gamma(\beta)} \right)^{\beta-1} \frac{1}{u} e^{-\frac{t}{u}} f(t) \, dt. \qquad (3.40)$$

We shall give some properties of the above operator.

3.3.1 Properties of beta-Sumudu transform

The proposed operator satisfies the following properties:

1. $$S_\beta(^A_0 D_t^\beta t^n)(u) = (n-1)! u^{n-1}. \qquad (3.41)$$

2. $$S_\beta(^A_0 D_t^\beta f \circ g)(u) = M(u)N(u). \qquad (3.42)$$

with $M(u)$ and $N(u)$ the Sumudu transform of f and g, respectively, with in addition $f * g$ differentiable.

3. Let $G(u)$ be the Sumudu transform of g, then

$$S_\beta(^A_0 D_t^\beta (f^{n-1}(t)))(u) = \frac{G(u)}{u^n} - \sum_{k=0}^{n-1} \frac{f^k(0)}{u^{n-k}}. \qquad (3.43)$$

4. $$S_\beta(^A_0 D_t^\beta D^{-V}(f(t)))(u) = (n-1)! u^{1-V} G(u), \quad Re[V] > 0. \qquad (3.44)$$

5. $$S_\beta(^A_0 D_t^\beta D^{-V}(t^{m-1} e^{at}))(u) = \Gamma(m) \frac{u^{m+2}}{(1-au)^{m-1}}. \qquad (3.45)$$

6. $$S_\beta(^A_0 D_t^\beta D^{-V}(e^{at}))(u) = \frac{u^V}{1-au}. \qquad (3.46)$$

We will prove the above properties case by case, starting with 1.

Proof. Using the definition of the beta-Sumudu transform, we have

$$S_\beta(^A_0 D_t^\beta t^n)(u) = \int_0^\infty \left(t + \frac{1}{\Gamma(\beta)} \right)^{\beta-1} \frac{1}{u} e^{-\frac{t}{u}} (^A_0 D_t^\beta t^n) \, dt \qquad (3.47)$$

Let $f(t) = t^n$, then

$$^A_0 D_t^\beta t^n = {}^A_0 D_t^\beta t^n = \lim_{\varepsilon \to 0} \frac{f\left(t + \varepsilon \left(t + \frac{1}{\Gamma(\beta)} \right)^{1-\beta} \right) - f(t)}{\varepsilon}. \qquad (3.48)$$

If we let $h = \varepsilon \left(t + \frac{1}{\Gamma(\beta)} \right)^{1-\beta}$ and using the fact that the function is differentiable, we obtain the following:

$$
{}_0^A D_t^\beta t^n = {}_0^A D_t^\beta t^n = \left(t + \frac{1}{\Gamma(\beta)} \right)^{1-\beta} \lim_{\varepsilon \to 0} \frac{f(t+h) - f(t)}{h}. \tag{3.49}
$$

Now replacing Equation (3.47) in Equation (3.49), we obtain the following:

$$
S_\beta ({}_0^A D_t^\beta t^n)(u) = \int_0^\infty \left(t + \frac{1}{\Gamma(\beta)} \right)^{\beta-1} \frac{1}{u} e^{-\frac{t}{u}}
$$
$$
\left(\left(t + \frac{1}{\Gamma(\beta)} \right)^{1-\beta} \lim_{h \to 0} \frac{f(t+h) - f(t)}{h} \right) dt. \tag{3.50}
$$

Rearranging, we obtain the following simplified version:

$$
S_\beta ({}_0^A D_t^\beta t^n)(u) = \int_0^\infty \frac{1}{u} e^{-\frac{t}{u}} \left(\lim_{h \to 0} \frac{f(t+h) - f(t)}{h} \right) dt
$$
$$
= S(t^{n-1})(u), \tag{3.51}
$$

where S is the Sumudu transform and is defined as:

$$
S(f(x))(u) = \int_0^\infty \frac{1}{u} e^{-\frac{t}{u}} f(t) \, dt. \tag{3.52}
$$

However, using the Sumudu transform properties, we have the following:

$$
S(t^{n-1})(u) = (n-1)! u^{n-1} \tag{3.53}
$$

and then

$$
S_\beta ({}_0^A D_t^\beta t^n)(u) = (n-1)! u^{n-1}. \tag{3.54}
$$

This completes the proof.   □

Proof. Let $M(u)$ and $N(u)$ be the Sumudu transform of f and g, respectively, and let $r = f * g$:

$$
S_\beta ({}_0^A D_t^\beta r(t))(u) = \int_0^\infty \frac{1}{u} e^{-\frac{t}{u}} \left(\lim_{h \to 0} \frac{r(t+h) - r(t)}{h} \right) dt
$$
$$
= \int_0^\infty \frac{1}{u} e^{-\frac{t}{u}} r' \, dt. \tag{3.55}
$$

Using the properties of the Sumudu transform, we obtain

$$S_\beta ({}_0^A D_t^\beta r(t))(u) = \frac{S(f * g)}{u} - f * g(0) = M(u)N(u) - (f * g)(0). \quad (3.56)$$

However, we know that $(f * g)(0) = 0$, therefore,

$$S_\beta ({}_0^A D_t^\beta r(t))(u) = M(u)N(u). \quad (3.57)$$

This completes the proof. □

Proof. For property 3, using the definition, we obtain

$$S_\beta ({}_0^A D_t^\beta f^{n-1}(t))(u) = \int_0^\infty \left(t + \frac{1}{\Gamma(\beta)} \right)^{\beta-1} \frac{1}{u} \times e^{-\frac{t}{u}}$$
$$\left(\left(t + \frac{1}{\Gamma(\beta)} \right)^{1-\beta} \lim_{h \to 0} \frac{f^{n-1}(t+h) - f^{n-1}(t)}{h} \right) dt.$$
$$(3.58)$$

After simplification, we obtain the following formula:

$$S_\beta ({}_0^A D_t^\beta f^{n-1}(t))(u) = \int_0^\infty \frac{1}{u} e^{-\frac{t}{u}} \left(\lim_{h \to 0} \frac{f^{n-1}(t+h) - f^{n-1}(t)}{h} \right) dt$$
$$= S(f^n(t)). \quad (3.59)$$

Using some properties of the Sumudu transform, we obtain

$$S_\beta ({}_0^A D_t^\beta f^{n-1}(t))(u) = \frac{G(u)}{u^n} - \sum_{k=0}^{n-1} \frac{f^k(0)}{u^{n-k}}. \quad (3.60)$$

And the required result is obtained. □

Proof. For fractional order derivatives, we have the following:

$$S_\beta ({}_0^A D_t^\beta D^V f(t))(u) = \int_0^\infty \left(t + \frac{1}{\Gamma(\beta)} \right)^{\beta-1} \frac{1}{u} \times e^{-\frac{t}{u}}$$
$$\left(\left(t + \frac{1}{\Gamma(\beta)} \right)^{1-\beta} \lim_{h \to 0} \frac{D^{-V}f(t+h) - D^{-V}f(t)}{h} \right) dt,$$
$$(3.61)$$

where D^{-V} is either the Caputo or Liouville–Riemann fractional order derivative. Thus after simplification, Equation (3.61) becomes:

$$S_\beta \left({}_0^A D_t^\beta D^V f(t) \right)(u) = \int_0^\infty \frac{1}{u} e^{-\frac{t}{u}} \left(\lim_{h \to 0} \frac{D^{-V}f(t+h) - D^{-V}f(t)}{h} \right) dt.$$
$$(3.62)$$

Now, using the properties of the Sumudu transform Equation (3.62) becomes:

$$S_\beta(^A_0 D^\beta_t D^{-V} f(t))(u) = u^{1-V} G(u) - D^{-V}(f(t))|_{t=0} = u^{1-V} G(U). \quad (3.63)$$

$\square$

Proof. For property 5, we have by using the definition and letting $f(t) = e^{at} t^{n-1}$, then

$$S_\beta(^A_0 D^\beta_t D^{-V} f(t))(u) = \int_0^\infty \frac{1}{u} e^{-\frac{t}{u}} \left(\lim_{h \to 0} \frac{D^{-V} f(t+h) - D^{-V} f(t)}{h} \right) dt$$

$$= u^{1-V} G(U) = \Gamma(m) \frac{u^{m+1}}{(1-au)^{m-1}}. \quad (3.64)$$

This is the completion of the proof. $\square$

3.4 DEFINITION AND PROPERTIES OF BETA-FOURIER TRANSFORM

In order to fit the concept of Fourier transform into the beta-calculus, we present in this section the definition of some useful properties of the β-Fourier transform.

Definition 3.4.1. Let f be a function defined in $(-\infty, \infty)$, then, we define the beta-Fourier transform of f as:

$$\mathcal{F}_\beta(f(x))(u) = \int_{-\infty}^\infty f(t) \left(t + \frac{1}{\Gamma(\beta)} \right)^{\beta-1} e^{-2\pi i t u} \, dt, \text{ for any real number } u. \quad (3.65)$$

When the independent variable t represents time, the transform variable u represents frequency.

3.4.1 Properties of the beta-Fourier transform
Here, we assume $f(x)$, $g(x)$, and $h(x)$ are integrable functions, are Lebesgue measurable on the real line, and satisfy

$$\int_{-\infty}^\infty |f(x)| < \infty, \quad (3.66)$$

then, the beta-Fourier transform has the following basic properties:

1. Linearity: For any complex numbers a and b, if $h(x) = af(x) + bg(x)$, then,

$$\mathcal{F}_\beta(h(t))(u) = a \cdot \mathcal{F}_\beta(f(t))(u) + b \cdot \mathcal{F}_\beta(g(t))(u). \tag{3.67}$$

2. Translation: For any real number x_0, if $h(x) = f(x - x_0)$, then,

$$\mathcal{F}_\beta(h(t)) = e^{-2i\pi x_0 u} \mathcal{F}_\beta(f). \tag{3.68}$$

3. Modulation: For any real number u_0, if $h(x) = e^{2i\pi x u_0} f(x)$, then,

$$\mathcal{F}_\beta(h(t))(u) = \mathcal{F}_\beta(u - u_0). \tag{3.69}$$

4. Scaling: For a nonzero real number a, if $h(x) = f(ax)$, then

$$\mathcal{F}_\beta h(x)(u) = \frac{1}{|a|} \mathcal{F}_\beta\left(\frac{u}{|a|}\right). \tag{3.70}$$

The case $a = -1$ leads to the time-reversal property, which states: if $h(x) = f(-x)$:

$$\mathcal{F}_\beta(h(t))(u) = \mathcal{F}_\beta(f)(-u). \tag{3.71}$$

5. Conjugation: If $h(x) = \overline{f(x)}$, then

$$\mathcal{F}_\beta(h)(u) = \mathcal{F}_\beta(\overline{f(-u)}). \tag{3.72}$$

6. Integration: substituting $u = 0$ in the definition, we obtain

$$F(f)(0) = \int_{-\infty}^{\infty} \left(x + \frac{1}{\Gamma(\beta)}\right)^{\beta-1} f(x)\, dx. \tag{3.73}$$

That is, the evaluation of the Fourier transform in the origin ($u = 0$) equals the beta-integral of f all over its domain.

7. Derivative:

$$\mathcal{F}_\beta\left(\frac{d^n f(x)}{dx^n}\right)(u) = \mathcal{F}_\beta(f)(u)(2i\pi \cdot u)^n \tag{3.74}$$

8. Multiplication with distance:

$$\mathcal{F}_\beta({}^A_0 D^\beta_x(x^n f(x)))(u) = \left(\frac{i}{2\pi}\right)^{n+1} \frac{d^{n+1}\mathcal{F}_\beta(f)(u)}{du^{n+1}}. \tag{3.75}$$

9. Convolution:

$$\mathcal{F}_\beta({}^A_0 D^\beta_x(f * g))(u) = \frac{i}{2\pi}\left[\frac{d\mathcal{F}_\beta(f)}{du} \cdot \mathcal{F}_\beta(g)(u) + \frac{d\mathcal{F}_\beta(g)}{du} \cdot \mathcal{F}_\beta(f)(u)\right]. \tag{3.76}$$

10. Product:

$$\mathcal{F}_\beta({}^A_0 D^\beta_x f(x) g(x))(u) = \frac{d(\mathcal{F}_\beta(f) * \mathcal{F}_\beta(g))(u)}{du}. \tag{3.77}$$

Theorem 3.4.1. *Plancherel theorem and Parseval theorem for beta-calculus: let f(x) and g(x) be integrable, and let $\mathcal{F}_\beta(f)$ and $\mathcal{F}_\beta(g)$ their Fourier transform, respectively. If f(x) and g(x) are also square-integrable, then we have Parseval's formula [74]:*

$$\int_{-\infty}^{\infty} f(x)\overline{g(x)}\, dx = \int_{-\infty}^{\infty} (\mathcal{F}_\beta(f)(u)) \cdot \overline{\mathcal{F}_\beta(g)(u)}\, du, \qquad (3.78)$$

where the bar denotes complex conjugation. The Plancherel theorem, which follows from the above, states that:

$$\int_{-\infty}^{\infty} |f(x)|^2\, dx = \int_{-\infty}^{\infty} |\mathcal{F}_\beta(f)(u)|^2\, du. \qquad (3.79)$$

The above properties are very easy to verify; we shall show the proof of Theorem 3.4.1.

Proof. In other words, let $f(x)$ be a function that is sufficiently smooth and that decays sufficiently quickly near infinity so that its integrals exist. Further, let $f(t)$ and f_v be beta-Fourier transform pairs so that:

$$f(x) = \int_{-\infty}^{\infty} f_v\, e^{-2\pi i v x}\, dv \qquad (3.80)$$

$$\bar{f}(x) = \int_{-\infty}^{\infty} f_{v'}\, e^{-2\pi i v' x}\, dv', \qquad (3.81)$$

where $\bar{z}$ denotes the complex conjugate. Then

$$
\begin{aligned}
\int_{-\infty}^{\infty} f_v(x)\bar{f}_{v'}(x)\, dx &= \int_{-\infty}^{\infty} \left[\int_{-\infty}^{\infty} f_v\, e^{-2\pi i v x}\, dv \cdot \int_{-\infty}^{\infty} f_{v'}\, e^{-2\pi i v' x}\, dv' \right] dt \\
&= \int_{-\infty}^{\infty} \int_{-\infty}^{\infty} \int_{-\infty}^{\infty} f_v \bar{f}_{v'}\, e^{2\pi i x(v'-v)}\, dv\, dv'\, dt \\
&= \int_{-\infty}^{\infty} \int_{-\infty}^{\infty} f_v \bar{f}_{v'} \delta(v'-v)\, dv\, dv' = \int_{-\infty}^{\infty} f_v f_{v'}\, dv \\
&= \int_{-\infty}^{\infty} |f_v|^2\, dv, \qquad (3.82)
\end{aligned}
$$

where $\delta(x - x_0)$ is the delta dirac function. This completes the proof. □

CHAPTER 4

Method for partial differential equations with beta-derivative

4.1 INTRODUCTION

The real problem with linear or nonlinear equation is to find a suitable analytical method that can be used to derive their exact or special solutions. It is no wonder that many scholars have devoted their attention in developing methods to handle these equations. Several methods were proposed; for instance, the Laplace transform method [75–77], the Mellin transform method [78], the Fourier transform method [79, 80], the Sumudu transform method [81–83], and the Green function method [84] for linear cases. The perturbation method [85], variational iteration method [86–88], homotopy decomposition and perturbation method [89–92], and others were developed for both linear and nonlinear cases. On the other hand, we should mention that mathematical models are a simplified description of physical reality expressed in mathematical terms. Thus, the investigation of the exact or approximate solution helps us to understand the means of these mathematical models. In most cases, it is difficult, or infeasible, to find the analytical solution, but a good numerical solution of the problems can be obtained. Numerical solutions or approximate analytical solutions become necessary. Numerical methods typically yield approximate solutions to the governing equation through the discretization of space and time, and can relax the rigid idealized conditions of analytical models or lumped-parameter models. They can, therefore, be more realistic and flexible for simulating field conditions. Within the discredited problem domain, the variable internal properties, boundaries, and stresses of the system are approximated. The main aim of this section is to present some iterative and numerical methods that will be used to solve ordinary and partial differential equation with beta-derivative. This will be presented in details in the next section, starting with iterative methods.

4.2 HOMOTOPY DECOMPOSITION METHOD

The homotopy decomposition method is the coupling of the Cauchy formula of n-integral together with the concept of homotopy. The method was first

Derivative with a New Parameter. http://dx.doi.org/10.1016/B978-0-08-100644-3.00004-0
© 2016 Elsevier Ltd. All rights reserved.

used to solve the groundwater flow equation and second to handle the time-fractional coupled-Korteweg–de Vries equations [93, 94]. To illustrate the basic idea of this method, we consider a general nonlinear nonhomogeneous fractional partial differential equation with initial conditions of the following form:

$$\,_0^A D_t^\beta (U(x,t)) = L(U(x,t)) + N(U(x,t)) + f(x,t), \quad 1 \le \beta \le 1. \tag{4.1}$$

Subject to the initial condition

$$U(x,0) = g(x), \tag{4.2}$$

where $\,_0^A D_t^\beta$ denotes the beta-derivative operator, f is a known function, N is the general nonlinear fractional differential operator, and L represents a linear fractional differential operator. The method first step here is to transform the fractional partial differential equation to the fractional partial integral equation by applying the inverse operator $\,_0^A I_t^\beta$ of both sides of Equation (4.1) to obtain the following

$$U(x,t) - U(x,0) = \int_0^x \left(y + \frac{1}{\Gamma(\beta)} \right)^{\beta-1} (L(U(x,y)) + N(U(x,y)) + f(x,y)) dy. \tag{4.3}$$

In the homotopy decomposition method, the basic assumption is that the solutions can be written as a power series with an introduction of an embedding parameter p

$$U(x,t,p) = \sum_{n=0}^{\infty} p^n U_n(x,t), \quad U(x,t) = \lim_{p \to 1} U(x,t,p), \tag{4.4}$$

and the nonlinear term can be decomposed as

$$N(U(x,t)) = \sum_{n=0}^{\infty} p^n \mathcal{H}_n(U), \tag{4.5}$$

where $p \in (0,1]$ is an embedding parameter, $\mathcal{H}_n(U)$ is a polynomial that can be generated by

$$\mathcal{H}_n(U_0, U_1, U_2 \ldots, U_n) = \frac{1}{n!} \frac{\partial^n}{\partial p^n} \left[N \left(\sum_{j=0}^{\infty} p^j U_j(x,t) \right) \right], \quad n = 0,1,2,\ldots. \tag{4.6}$$

However, replacing Equations (4.5), (4.4), and (4.6) into Equation (4.3), we obtain

$$\sum_{n=0}^{\infty} p^n U_n(x,t) - U(x,0) = p \int_0^x \left(y + \frac{1}{\Gamma(\beta)}\right)^{\beta-1} \left(L\left(\sum_{n=0}^{\infty} p^n U_n(x,y)\right)\right.$$

$$\left. + N\left(\sum_{n=0}^{\infty} p^n U_n(x,y)\right) + f(x,y)\right) dy. \qquad (4.7)$$

Comparison of the terms of same powers of p gives solutions of various orders with the first term

$$U_0(x,t) = U(x,0). \qquad (4.8)$$

One of the important parts of any iteration method is to prove the uniqueness and the convergence of the method; we are going to show the analysis underpinning the convergence and the uniqueness of the proposed method for the general solution when $p = 1$.

Theorem 4.2.1. *Assume that X and Y are Banach spaces and $V : X \longrightarrow Y$ is contraction nonlinear mapping. If the progression engendered by the three-dimensional homotopy decomposition method is regarded as*

$$U_n(x,t) = V(U_{n-1}(x,t)) = \sum_{k=0}^{n-1} U_k(x,t), \quad n = 1,2,3,\ldots, \qquad (4.9)$$

then the following statements hold

(a) $\|U_n(x,t) - U(x,t)\| \le \rho^n \|T(x,t) - U(x,t)\|$, *with* $0 < \rho < 1$.
(b) *For any n greater than 0, $T(x,t)$ is always in the neighborhood of the exact solution $U(x,t)$.*
(c) $\lim_{n \to \infty} U_n(x,t) = T(x,t)$.

Proof. The proof of (a) shall be achieved via induction on the natural number n. However, when $n = 1$, we have the following

$$\|U_1(x,t) - U(x,t)\| = \|V(U_0(x,t)) - U(x,t)\|. \qquad (4.10)$$

However, by hypothesis, we have that V has a fixed point, which is the exact solution. Because if $U(x,t)$ is the exact solution, then,

$$U(x,t) = U_\infty(x,t) = V\left(\sum_{k=0}^{\infty-1} U_k(x,t)\right) = V\left(\sum_{k=0}^{\infty} U_k(x,t)\right) = \sum_{k=0}^{\infty} U_k(x,t),$$

$$(4.11)$$

since $\infty - 1$ is the same as ∞, therefore we have that

$$U(x, t) = V(U(x, t)).\qquad(4.12)$$

Then,

$$\|U_1(x, t) - U(x, t)\| = \|V(U_0(x, t)) - V(U(x, t))\|.\qquad(4.13)$$

Since V is a contractive nonlinear mapping, we shall have the following inequality

$$\|V(U_0(x, t)) - V(U(x, t))\| = \|V(U_{n-1}(x, t)) - T(x, t)\|$$
$$= V(T_{n-1}(x, t)) - V(T(x, t))\|.\qquad(4.14)$$

Using the fact that V is a nonlinear contractive mapping, we have the following

$$V(U_{n-1}(x, t)) - V(U(x, t))\| < \rho\|(U_{n-1}(x, t)) - V(U(x, t))\|.\qquad(4.15)$$

Furthermore, using the induction hypothesis, we arrive at

$$\rho\|U_{n-1}(x, t) - V(U(x, t))\| < \rho\rho^{n-1}\|U_0(x, t) - U(x, t)\|.\qquad(4.16)$$

And the proof is completed.   □

Proof. Again we shall prove this by employing induction technique on m. Now for $m = 0$, we have that

$$U_0(x, t) = T(x, t) = \sum_{w=0}^{n-1}\sum_{h=0}^{m-1}\frac{f_{w,h}}{w!h!}x^w t^h.\qquad(4.17)$$

According to the idea of the homotopy decomposition method, the above is the contribution of the initial conditions. More importantly, the above is nothing more than the Taylor series of the exact solution of order 1, thus this leads us to the situation that we can find a positive real number r such that

$$\|U_0(x, t) - U(x, t)\| < r.\qquad(4.18)$$

This is true, because the contribution of the initial conditions is in the same neighborhood of the exact solution. Then the property is verified for $m = 0$. Let us assume that the property is also true for $m - 1$; that is, we assume that we can find a positive real number r such that

$$\|U_{m-1}(x, t) - U(x, t)\| < r.\qquad(4.19)$$

We now want to show that the property is also true for m. In fact

$$\|U_m(x, y, t) - U(x, t)\| = \|V(U_{m-1}(x, t)) - V(U(x, t))\|.\qquad(4.20)$$

Using the fact that V is a nonlinear contractive mapping leads us to obtain

$$\|V(U_{m-1}(x,t)) - V(U(x,t))\| < \rho \|U_{m-1}(x,t) - U(x,t)\| < \rho r. \quad (4.21)$$

Since $\rho < 1$, we finally have

$$\|U_m(x,t) - U(x,t)\| < r, \quad (4.22)$$

and this completes the proof. □

Proof. The proof of (c) is directly achieved using (a) as follows

$$\lim_{n \to \infty} \|U_n(x,t) - U(x,t)\| \le \lim_{n \to \infty} \rho^n \|T(x,t) - U(x,t)\| = 0. \quad (4.23)$$

Then

$$\lim_{n \to \infty} U_n(x,t) = U(x,t). \quad (4.24)$$

We will illustrate the use of this technique by solving a simple equation. □

Example 4.1. Consider the following decay equation with beta-derivative

$$_0^A D_t^\beta (N(t)) = aN(t), \quad N(0) = N_0 \quad (4.25)$$

To solve Equation (4.25) using the described technique, we apply on both sides the beta-integral to obtain

$$N(t) - N(0) = {}_0^A I_t^\beta (aN(t)), \quad (4.26)$$

where we further assume that

$$N(t) = \lim_{p \to 1} \sum_{i=0}^{\infty} N_i(t) p^i. \quad (4.27)$$

Replacing Equation (4.27) into Equation (4.25), and comparing terms of the same power of p, we obtain the following beta-integral equations

$$N_0(t) = N_0$$
$$N_1(t) = {}_0^A I_t^\beta (N_0(t))$$
$$\vdots$$
$$N_n(t) = {}_0^A I_t^\beta (N_{n-1}(t)). \quad (4.28)$$

Now integrating the above yields

$$N_0(t) = N_0$$

$$N_1(t) = N_0 \left[\frac{a}{\beta} \left(\frac{1}{\Gamma^\beta(\beta)} - \left(t + \frac{1}{\Gamma(\beta)} \right)^\beta \right) \right]$$

$$N_2(t) = \frac{1}{2!} N_0 \left[\frac{a}{\beta} \left(\frac{1}{\Gamma^\beta(\beta)} - \left(t + \frac{1}{\Gamma(\beta)} \right)^\beta \right) \right]^2$$

$$N_3(t) = \frac{1}{3!} N_0 \left[\frac{a}{\beta} \left(\frac{1}{\Gamma^\beta(\beta)} - \left(t + \frac{1}{\Gamma(\beta)} \right)^\beta \right) \right]^3$$

$$\vdots$$

$$N_n(t) = \frac{1}{n!} N_0 \left[\frac{a}{\beta} \left(\frac{1}{\Gamma^\beta(\beta)} - \left(t + \frac{1}{\Gamma(\beta)} \right)^\beta \right) \right]^n. \tag{4.29}$$

Therefore,

$$N(t) = \sum_{n=0}^{\infty} N_n = \sum_{n=0}^{\infty} \frac{1}{n!} N_0 \left[\frac{a}{\beta} \left(\frac{1}{\Gamma^\beta(\beta)} - \left(t + \frac{1}{\Gamma(\beta)} \right)^\beta \right) \right]^n$$

$$= N_0 \, e^{\left(\left[\frac{a}{\beta} \left(\frac{1}{\Gamma^\beta(\beta)} - \left(t + \frac{1}{\Gamma(\beta)} \right)^\beta \right) \right] \right)}. \tag{4.30}$$

This is the exact solution of our example.

Example 4.2. We consider the one-dimensional fractional wave-like equation

$$_0^A D_t^\beta u(x,t) = \frac{1}{2} x^2 u_{xx}, \quad 0 < x < 1, \ t > 0, \ u(x,0) = x^2. \tag{4.31}$$

To solve Equation (4.31), we apply the beta-integral on both sides of Equation (4.31) to obtain

$$u(x,t) - u(x,0) = {_0^A D_t^\beta} \left(\frac{1}{2} x^2 u_{xx} \right), \tag{4.32}$$

assuming that the exact solution of Equation (4.65) is in the form of series as follows

$$u(x,t) = \lim_{p \to 1} \sum_{n=0}^{\infty} u_n(x,t) p^n. \tag{4.33}$$

Now replacing Equation (4.33) into Equation (4.65), comparing terms of the same power of p, we obtain the following beta-integral equations

$$u_0(x, t) = x^2$$

$$u_1(x, t) = {}_0^A I_t^\beta \left(\frac{1}{2} x^2 u_{0(xx)} \right)$$

$$u_2(x, t) = {}_0^A I_t^\beta \left(\frac{1}{2} x^2 u_{1(xx)} \right)$$

$$\vdots$$

$$u_{n+1}(x, t) = {}_0^A I_t^\beta \left(\frac{1}{2} x^2 u_{n(xx)} \right). \tag{4.34}$$

Now integrating the above yields

$$u_0(x, t) = x^2$$

$$u_1(x, t) = x^2 \frac{\left(t + \frac{1}{\Gamma(\beta)} \right)^\beta}{1!(\beta)}$$

$$u_2(x, t) = x^2 \frac{\left(\left(t + \frac{1}{\Gamma(\beta)} \right)^\beta \right)^2}{2!(\beta)^2}$$

$$u_3(x, t) = x^2 \frac{\left(\left(t + \frac{1}{\Gamma(\beta)} \right)^\beta \right)^3}{3!(\beta)^3}$$

$$\vdots$$

$$u_{n+1}(x, t) = x^2 \frac{\left(\left(t + \frac{1}{\Gamma(\beta)} \right)^\beta \right)^n}{n!(\beta)^n}. \tag{4.35}$$

Therefore, the exact solution of Equation (4.65) is provided as

$$u(x, t) = \sum_{n=0}^{\infty} u_n(x, t) = \sum_{n=0}^{\infty} \frac{\left(\left(t + \frac{1}{\Gamma(\beta)} \right)^\beta \right)^n}{n!(2 - \beta)^n} = x^2 e^{\left(\frac{\left(t + \frac{1}{\Gamma(\beta)} \right)^\beta - \left(\frac{1}{\Gamma(\beta)} \right)^\beta}{\beta} \right)}. \tag{4.36}$$

This indeed is the exact solution of Equation (4.65).

Example 4.3. Consider the nonlinear wave-like equation with variable coefficients

$$_0^A D_t^{2\beta} u = x^2 \frac{\partial u_x u_{xx}}{\partial x} - x^2 U_{xx}^2 - u, \quad 0 < x < 1, \quad t > 0, \quad 0 \le \beta \le 1. \quad (4.37)$$

This is subjected to the initial conditions

$$u(x,0) = 0, \quad u_t(x,0) = x^2. \quad (4.38)$$

Applying the same routine like before, we obtain the following integral equation

$$u_0(x,t) = tu_t(x,0) = tx^2$$

$$u_1(x,t) = {}_0^A I_t^\beta \left(x^2 \frac{\partial u_{0x} u_{0xx}}{\partial x} - x^2 u_{0xx}^2 - u_0 \right)$$

$$u_2(x,t) = {}_0^A I_t^\beta \left(2x^2 \frac{\partial u_{1,x} u_{0,xx}}{\partial x} - 2x^2 u_{1,xx} u_{0,xx} - u_0 \right)$$

$$\vdots$$

$$u_{n+1}(x,t) = {}_0^A I_t^\beta \left(2x^2 \frac{\partial}{\partial x} \left(\sum_{j=0}^n u_{n-j,x} u_{j,xx} \right) - 2x^2 \left(\sum_{j=0}^n u_{n-j,xx} u_{j,xx} \right) - u_n \right).$$

$$(4.39)$$

The following solutions are obtained:

$$u_0(x,t) = tx^2$$

$$u_1(x,t) = x^2(-1)^1 \frac{\left(t + \frac{1}{\Gamma(\beta)} \right)^{3(\beta)}}{3!(\beta)}$$

$$u_2(x,t) = x^2(-1)^2 \frac{\left(\left(t + \frac{1}{\Gamma(\beta)} \right)^{2-\beta} \right)^5}{5!(\beta)^5}$$

$$u_3(x,t) = (-1)^3 \frac{\left(\left(t + \frac{1}{\Gamma(\beta)} \right)^\beta \right)^7}{7!(\beta)^7}$$

$$\vdots$$

$$u_{n+1}(x,t) = (-1)^n \frac{\left(\left(t + \frac{1}{\Gamma(\beta)} \right)^\beta \right)^{2n+1}}{(2n+1)!(\beta)^{2n+1}}. \quad (4.40)$$

Therefore, the exact solution of Equation (4.37) is provided by

$$u(x, t) = a \sum_{n=0}^{\infty} (-1)^n \frac{\left(\left(t + \frac{1}{\Gamma(\beta)}\right)^{\beta}\right)^{2n+1}}{(2n+1)!(\beta)^{2n+1}}. \tag{4.41}$$

However, applying the initial condition, we obtain

$$u(x, t) = x^2 \sum_{n=0}^{\infty} (-1)^n \frac{\left(\left(t + \frac{1}{\Gamma(\beta)}\right)^{\beta} - \left(\frac{1}{\Gamma(\beta)}\right)^{\beta}\right)^{2n+1}}{(2n+1)!(2-\beta)^{2n+1}} = x^2 \sin^{\beta}(t).$$

$$\tag{4.42}$$

This is the exact solution of our equation.

4.3 VARIATIONAL ITERATION METHOD

Very recently, it was recognized that the variational iteration method [95, 96] can be an effective procedure for the solution of various nonlinear problems without usual restrictive assumptions. The method, extensively worked out by numerous authors, has been maturing into a fully fledged theory, more and more merits have been discovered, and some modifications have been suggested to overcome the demerit arising in the solution procedure. Applications of the method have been enlarged due to its flexibility, convenience, and accuracy. A guided tour through the mathematics needed for a proper understanding of the variational iteration method as applied to various nonlinear problems is available [97], for a relatively comprehensive survey on the method and its applications. The variational iteration method has been favorably applied to various kinds of nonlinear problems. The main property of the method is in its flexibility and ability to solve nonlinear equations accurately and conveniently. In this section, a presentation of recent trends and developments in the use of the method within the scope of beta-differential equations is made.

4.3.1 Methodology and stability analysis

To illustrate the basic idea of the method, we consider the general nonlinear beta-ordinary differential equation:

$$_0^A D_t^{\beta} u(t) + L(u(t)) + N(u(t)) = g(t), \tag{4.43}$$

where L is a linear operator, N is a nonlinear operator, and $g(t)$ is a given continuous function. The basic character of the method is to construct a functional for the system, which reads

$$u_{n+1}(x) = u_n(x) + {}_0^A I_t^\beta \lambda(s) \left[{}_0^A D_s^\beta (u_n) + L[u_n(s)] + N[\bar{u}_n(s)] - g(s) \right],$$

(4.44)

where λ is the Lagrange multiplier which can be identified optimally by the mean of variational theory, u_n is the nth approximation solution, and $\bar{u}_n$ denotes a restricted variation, that is, $\delta \bar{u}_n = 0$. Here the stability analysis will be achieved using the concept of the Fredholm integral equation of second kind in the general case, which reads

$$u(x) = f(x) + \lambda \int_a^b k(x, t) u(t) \, dt,$$

(4.45)

with $k(x, t)$ being the kernel of the integral equation. There is a simple associate recursive formula for Equation (4.45) in the form

$$u_n(x) = f(x) + \lambda \int_a^b k(x, t) u_n(t) \, dt.$$

(4.46)

Now, we show that the nonlinear mapping T_β, defined by

$$u_{n+1}(x) = T_\beta(u_n(x)) = f(x) + \lambda \int_a^b k(x, t) u_n(t) \, dt,$$

(4.47)

is T_β-stable in $L^2[a, b]$. Note: Let $(X, \|.\|)$ be a Banach space and T_β a self-map of X. Let $x_{n+1} = f(T_\beta, x_n)$ be some iteration procedure. Suppose that $F(T_\beta)$, the fixed point set of T_β, is nonempty and that x_n converges to a point $q \in F(T_\beta)$. Let $y_n \subset X$ and define $e_n = \|y_{n+1} - f(T_\beta, y_n)\|$. If $\lim e_n = 0$ implies that $\lim y_n = q$, then the iteration procedure $x_{n+1} = f(T_\beta, x_n)$ is said to be T_β-stable. However, without loss of generality, we may assume that y_n is bounded, otherwise it cannot possible converge. If these conditions hold for $x_{n+1} = T_\beta x_n$, that is, Picard's iteration, then we will say that Picard's iteration is T_β-stable.

Theorem 4.3.1 (See [98]). *Let $(X, \|.\|)$ be a Banach space and T_β a self-map of X satisfying*

$$\|T_\beta x - T_\beta y\| \leq L \|x - T_\beta y\| + a \|x - y\|,$$

(4.48)

for all $x, y \in X$, where $L \geq 0$, $0 \leq \alpha < 1$. Suppose that T_β has a fixed point p. Then, T_β is Picard T_β-stable.

With this information in hand, we first show that the nonlinear mapping T_β has a fixed point. For $n, m \in \mathbb{N}$, we have

$$
\begin{aligned}
\|T(u_m(x)) - T(u_n(x))\| &= \|u_{m+1}(x) - u_{n+1}(x)\| \\
&= \left\| \lambda_0^A I_t^\beta k(x, t)(u_m(t) - u_n(t)) \right\| |\lambda| \left[{}_0^A I_t^\beta \left({}_0^A I_t^\beta k^2(x, t) \right) \right]^{1/2} \\
&\quad \times \|u_n(x) - u_m(x)\|.
\end{aligned} \tag{4.49}
$$

Nevertheless, if

$$
\|\lambda\| \left[{}_0^A I_t^\beta \left({}_0^A I_t^\beta k^2(x, t) \right) \right]^{-1/2}, \tag{4.50}
$$

then the nonlinear mapping T_β has a fixed point. Secondly, we show that the nonlinear mapping T_β satisfies Equation (4.48). Let Equation (4.46) hold. Thus, putting $L = 0$ and $\alpha = |\lambda[{}_0^A I_t^\beta ({}_0^A I_t^\beta k^2(x, t))]^{1/2}$ shows that Equation (4.48) holds for the nonlinear mapping T_β. All of the conditions of Theorem 4.3.1 hold for nonlinear mapping T_β and hence it is T_β-stable. As result, we can state the following theorem.

Theorem 4.3.2. *Using the iteration scheme*

$$
u_0(x) = f(x),
$$

$$
u_{n+1}(x) = T_\beta(u_n(x)) = f(x) + \lambda \int_a^b k(x, t) u_n(t)\, dt, \tag{4.51}
$$

for $n = 0, 1, 3, \ldots$ to construct a sequence of successive iterations $[u_n(x)]$ to the solution of Equation (4.46). In addition, if

$$
\|\lambda\| \left[{}_0^A I_t^\beta \left({}_0^A I_t^\beta k^2(x, t) \right) \right]^{-1/2}, \tag{4.52}
$$

$L = 0$ and $\alpha = |\lambda[{}_0^A I_t^\beta ({}_0^A I_t^\beta k^2(x, t))]^{-1/2}$, then the nonlinear mapping T_β, in the norm of $L^2(a, b)$ is T_β-stable.

Theorem 4.3.3 (See [99]). *Use the iteration scheme*

$$
u_0(x) = f(x),
$$

$$
u_{n+1}(x) = f(x) + \lambda \int_a^b k(x, t) u_n(t)\, dt, \tag{4.53}
$$

for $n = 0, 1, 3, \ldots$ to construct a sequence of successive iterations $[u_n(x)]$ to the solution of Equation (4.46). In addition, let

$$
\left({}_0^A I_t^\beta \left({}_0^A I_t^\beta k^2(x, t) \right) \right) = B^2 < \infty, \tag{4.54}
$$

and assume that $f(x) \in L^2(a,b)$, then, if $\lambda < \frac{1}{B}$, the above iteration converges, in the norm $L^2(a,b)$ to solution of Equation (4.46).

Corollary 4.3.1. *Consider the iteration scheme*

$$u_0(x) = f(x),$$

$$u_{n+1}(x) = T_\beta(u_n(x)) = f(x) + \lambda \int_a^b k(x,t)u_n(t)\, dt, \qquad (4.55)$$

for $n = 0,1,2,\ldots$ *if* $L = 0$ *and* $\alpha = |\lambda[_0^A I_t^\beta (_0^A I_t^\beta k^2(x,t))]^{-1/2}$, *then stability of the nonlinear mapping* T_β *in norm of* $L^2(a,b)$ *is a coefficient condition for the above iterative to converge in the norm of* $L^2(a,b)$, *and to the solution of Equation (4.46).*

We shall illustrate the use of this method with the following examples.

Example 4.4. Consider the following decay equation we get the beta-derivative

$$_0^A D_t^\beta (N(t)) = aN(t), \quad N(0) = N_0 \qquad (4.56)$$

Now following the methodology of the variation iteration method, we first obtain the Lagrange multiplier and we obtain the following recursive formula

$$N_0(t) = N_0$$

$$N_1(t) = N_0 + N_0 \left[\frac{a}{\beta} \left(\frac{1}{\Gamma^\beta(\beta)} - \left(t + \frac{1}{\Gamma(\beta)}\right)^\beta \right) \right]$$

$$N_2(t) = N_0 + N_0 \left[\frac{a}{\beta} \left(\frac{1}{\Gamma^\beta(\beta)} - \left(t + \frac{1}{\Gamma(\beta)}\right)^\beta \right) \right]$$

$$+ \frac{1}{2!} N_0 \left[\frac{a}{\beta} \left(\frac{1}{\Gamma^\beta(\beta)} - \left(t + \frac{1}{\Gamma(\beta)}\right)^\beta \right) \right]^2$$

$$N_3(t) = N_0 + N_0 \left[\frac{a}{\beta} \left(\frac{1}{\Gamma^\beta(\beta)} - \left(t + \frac{1}{\Gamma(\beta)}\right)^\beta \right) \right]$$

$$+ \frac{1}{2!} N_0 \left[\frac{a}{\beta} \left(\frac{1}{\Gamma^\beta(\beta)} - \left(t + \frac{1}{\Gamma(\beta)}\right)^\beta \right) \right]^2$$

$$+ \frac{1}{3!} N_0 \left[\frac{a}{\beta} \left(\frac{1}{\Gamma^\beta(\beta)} - \left(t + \frac{1}{\Gamma(\beta)} \right)^\beta \right) \right]^3$$

$$\vdots$$

$$N_n(t) = N_0 + N_0 \left[\frac{a}{\beta} \left(\frac{1}{\Gamma^\beta(\beta)} - \left(t + \frac{1}{\Gamma(\beta)} \right)^\beta \right) \right]$$

$$+ \frac{1}{2!} N_0 \left[\frac{a}{\beta} \left(\frac{1}{\Gamma^\beta(\beta)} - \left(t + \frac{1}{\Gamma(\beta)} \right)^\beta \right) \right]^2$$

$$+ \frac{1}{3!} N_0 \left[\frac{a}{\beta} \left(\frac{1}{\Gamma^{2-\beta}(\beta)} - \left(t + \frac{1}{\Gamma(\beta)} \right)^\beta \right) \right]^3$$

$$+ \cdots + \frac{1}{n!} N_0 \left[\frac{a}{\beta} \left(\frac{1}{\Gamma^\beta(\beta)} - \left(t + \frac{1}{\Gamma(\beta)} \right)^\beta \right) \right]^n. \tag{4.57}$$

In this method, the approximate solution can be obtained by taking the limit as $n \to \infty$ means. To obtain the approximate solution we calculate

$$\lim_{n \to \infty} u_n = u(t). \tag{4.58}$$

Therefore,

$$\lim_{n \to \infty} u_{n+1}(t) = u(t) = \sum_{n=0}^{\infty} \frac{1}{n!} N_0 \left[\frac{a}{\beta} \left(\frac{1}{\Gamma^\beta(\beta)} - \left(t + \frac{1}{\Gamma(\beta)} \right)^\beta \right) \right]^n$$

$$= N_0 \, e^{\left[\frac{a}{\beta} \left(\frac{1}{\Gamma^\beta(\beta)} - \left(t + \frac{1}{\Gamma(\beta)} \right)^\beta \right) \right]}. \tag{4.59}$$

Note that if $\beta = 1$ we obtain the exact solution of our equation.

Example 4.5. Consider the nonlinear wave-like equation with variable coefficients

$$^A_0 D_t^{2\beta} u = x^2 \frac{\partial u_x u_{xx}}{\partial x} - x^2 U_{xx}^2 - u, \quad 0 < x < 1, \ t > 0, \ 0 \le \beta \le 1, \tag{4.60}$$

subjected to the initial conditions

$$u(x, 0) = 0, \quad u_t(x, 0) = x^2. \tag{4.61}$$

Applying the same routine as before, we obtain the following integral equation

$$u_0(x, t) = tx^2$$

$$u_1(x, t) = x^2(-1)^1 \frac{\left(t + \frac{1}{\Gamma(\beta)}\right)^{3(\beta)}}{3!(\beta)} + tx^2$$

$$u_2(x, t) = x^2(-1)^2 \frac{\left(\left(t + \frac{1}{\Gamma(\beta)}\right)^{\beta}\right)^5}{5!(\beta)^5} + x^2(-1)^1 \frac{\left(t + \frac{1}{\Gamma(\beta)}\right)^{3(\beta)}}{3!(\beta)} + tx^2$$

$$u_3(x, t) = (-1)^3 \frac{\left(\left(t + \frac{1}{\Gamma(\beta)}\right)^{\beta}\right)^7}{7!(\beta)^7} + x^2(-1)^2 \frac{\left(\left(t + \frac{1}{\Gamma(\beta)}\right)^{\beta}\right)^5}{5!(\beta)^5}$$

$$+ x^2(-1)^1 \frac{\left(t + \frac{1}{\Gamma(\beta)}\right)^{3(\beta)}}{3!(\beta)} + tx^2$$

$$\vdots$$

$$u_{n+1}(x, t) = (-1)^n \frac{\left(\left(t + \frac{1}{\Gamma(\beta)}\right)^{\beta}\right)^{2n+1}}{(2n + 1)!(\beta)^{2n+1}} + \cdots + (-1)^3 \frac{\left(\left(t + \frac{1}{\Gamma(\beta)}\right)^{\beta}\right)^7}{7!(\beta)^7}$$

$$+ x^2(-1)^2 \frac{\left(\left(t + \frac{1}{\Gamma(\beta)}\right)^{\beta}\right)^5}{5!(\beta)^5} + x^2(-1)^1 \frac{\left(t + \frac{1}{\Gamma(\beta)}\right)^{3(\beta)}}{3!(\beta)} + tx^2.$$

$$(4.62)$$

In this method, the approximate solution can be obtained by taking the limit as $n \to \infty$ means. To obtain the approximate solution we calculate

$$\lim_{n \to \infty} u_n = u(x, t). \qquad (4.63)$$

Therefore,

$$\lim_{n \to \infty} u_{n+1}(t) = u(x, t) = x^2 \sum_{n=0}^{\infty} (-1)^n \frac{\left(\left(t + \frac{1}{\Gamma(\beta)}\right)^{\beta} - \left(\frac{1}{\Gamma(\beta)}\right)^{\beta}\right)^{2n+1}}{(2n + 1)!(\beta)^{2n+1}}$$

$$= x^2 \sin^{\beta}(t). \qquad (4.64)$$

Note that if $\beta = 1$ we obtain the exact solution of our equation.

Example 4.6. We consider the one-dimensional fractional wave-like equation

$$_0^A D_t^\beta u(x,t) = \frac{1}{2}x^2 u_{xx}, \quad 0 < x < 1, \quad t > 0, \quad u(x,0) = x^2. \qquad (4.65)$$

Following the routine as previously, we obtain

$$u_0(x,t) = x^2$$

$$u_1(x,t) = x^2 + x^2 \frac{\left(t + \frac{1}{\Gamma(\beta)}\right)^\beta}{1!(\beta)}$$

$$u_2(x,t) = x^2 + x^2 \frac{\left(t + \frac{1}{\Gamma(\beta)}\right)^\beta}{1!(\beta)} + x^2 \frac{\left(\left(t + \frac{1}{\Gamma(\beta)}\right)^\beta\right)^2}{2!(\beta)^2}$$

$$u_3(x,t) = x^2 + x^2 \frac{\left(t + \frac{1}{\Gamma(\beta)}\right)^\beta}{1!(\beta)} + x^2 \frac{\left(\left(t + \frac{1}{\Gamma(\beta)}\right)^\beta\right)^2}{2!(\beta)^2} + x^2 \frac{\left(\left(t + \frac{1}{\Gamma(\beta)}\right)^\beta\right)^3}{3!(\beta)^3}$$

$$\vdots$$

$$u_{n+1}(x,t) = x^2 + x^2 \frac{\left(t + \frac{1}{\Gamma(\beta)}\right)^\beta}{1!(2-\beta)} + x^2 \frac{\left(\left(t + \frac{1}{\Gamma(\beta)}\right)^\beta\right)^2}{2!(\beta)^2}$$

$$+ \cdots + x^2 \frac{\left(\left(t + \frac{1}{\Gamma(\beta)}\right)^\beta\right)^n}{n!(\beta)^n}. \qquad (4.66)$$

Therefore, the exact solution of Equation (4.65) is provided as

$$u(x,t) = \lim_{n \to \infty} u_n(x,t) = \sum_{n=0}^{\infty} x^2 \frac{\left(\left(t + \frac{1}{\Gamma(\beta)}\right)^\beta\right)^n}{n!(\beta)^n} = x^2\, e^{\left(\frac{\left(t+\frac{1}{\Gamma(\beta)}\right)^\beta - \left(\frac{1}{\Gamma(\beta)}\right)^\beta}{\beta}\right)}.$$

$$(4.67)$$

This indeed is the exact solution of Equation (4.65).

4.4 SUMUDU DECOMPOSITION METHOD

The Sumudu transform operator is an integral transform similar to the Laplace transform method, introduced in the early 1990s by Watugala, as noted earlier, to solve differential equations and control engineering problems. However, with the difficulties faced while solving the nonlinear equations, researchers have combined the concept of iterative method together with the Sumudu transform operator. The method has been employed to solve many linear and nonlinear equation in the literature; see, for instance, [100]. To illustrate the basic idea of this method, we consider a general nonlinear nonhomogenous partial differential equation with the initial conditions of the form

$$
{}_0^A D_t^\beta(u(x,t)) + L(u(x,t)) + N(u(x,t)) = g(x,t), \quad u(x,0) = h(x), \quad u_t(x,0) = f(x).
$$
(4.68)

L is the linear differential operator, N represents the general nonlinear differential operator, and $g(x,t)$ is the source term. Taking the beta-Sumudu transform operator on both sides of Equation (4.89), we get

$$
S_\beta\left({}_0^A D_t^\beta(u(x,t))\right) + S_\beta(L(u(x,t))) + S_\beta(N(u(x,t))) = S_\beta(g(x,t))(u). \quad (4.69)
$$

Using the differentiation property of the Sumudu transform operator and above initial conditions, we have

$$
S_\beta(u(x,t))(u) = uS_\beta(g(x,t)) + uf(x) - uS_\beta(L(u(x,t))) - uS_\beta(N(u(x,t))).
$$
(4.70)

Now, applying the inverse Sumudu transform on both sides of Equation (4.91), we get

$$
u(x,t) = F(x,t) - S^{-1}[-uS_\beta(L(u(x,t))) - uS_\beta(N(u(x,t)))], \quad (4.71)
$$

where $F(x,t)$ represents the term arising from the source term and the prescribed initial conditions. Now, we apply the homotopy perturbation method

$$
u(x,t) = \lim_{p \to 1} \sum_{n=0}^{\infty} p^n u_n(x,t), \quad (4.72)
$$

and the nonlinear term can be decomposed as

$$
N(u(x,t)) = \sum_{n=0}^{\infty} p^n H_n(u), \quad (4.73)
$$

where

$$H_n(u_0, u_1, \ldots, u_n) = \frac{1}{n!} \frac{\partial^n}{\partial p^n} \left[N \left(\sum_{i=0}^{\infty} p^i u(x, t) \right) \right]. \qquad (4.74)$$

Replacing this into Equation (4.89)

$$\sum_{n=0}^{\infty} p^n u_n(x, t) = F(x, t) - pS^{-1} \left[-u S_\beta \left(L \left(\sum_{n=0}^{\infty} p^n u_n(x, t) \right) \right) \right.$$

$$\left. - u S_\beta \left(\sum_{n=0}^{\infty} p^n H_n(u) \right) \right]. \qquad (4.75)$$

After a comparison of coefficients of like power p, the following components are obtained

$$p^0 : u_0(x, t) = F(x, t)$$
$$p^1 : u_1(x, t) = -S^{-1} \left[u S_\beta [L u_0(x, t) + H_0(u_0)] \right]$$
$$p^2 : u_2(x, t) = -S^{-1} \left[u S_\beta [L u_1(x, t) + H_0(u_0, u_1)] \right]$$
$$p^3 : u_3(x, t) = -S^{-1} \left[u S_\beta [L u_2(x, t) + H_0(u_0, u_1, u_2)] \right]$$
$$\vdots$$
$$p^n : u_n(x, t) = -S^{-1} \left[u S_\beta [L u_{(n-1)}(x, t) + H_0(u_0, u_1, u_2, \ldots, u_{(n-1)})] \right]. \qquad (4.76)$$

We illustrate the above technique by solving some partial beta-equations.

Example 4.7. Consider the following two-dimensional initial boundary value problem describing the model of wave-like

$$u_{tt} = \frac{1}{12} (x^2 u_{xx} + y^2 u_{yy}), \quad 0 < x, y < 1, \quad t > 0, \qquad (4.77)$$

subject to the Neumann boundary conditions

$$u_x(0, y, t) = u_y(x, 0, t) = 0, \quad u_x(1, y, t) = 4 \cosh(t), \quad u_x(x, 1, t) = 4 \sinh(t), \qquad (4.78)$$

and initial conditions

$$u(x, y, 0) = x^4, \quad u_t(x, y, 0) = y^4. \qquad (4.79)$$

Using the routine presented in this section, we have

$$\sum_{n=0}^{\infty} p^n u_n(x,y,t) = x^4 + y^4 t$$

$$+ p \left(\frac{1}{12} x^2 S^{-1} \left[u^2 S_\beta \left(\sum_{n=0}^{\infty} p^n u_n(x,y,t) \right)_{xx} \right] \right)$$

$$+ p \left(\frac{1}{12} y^2 S^{-1} \left[u^2 S_\beta \left(\sum_{n=0}^{\infty} p^n u_n(x,y,t) \right)_{yy} \right] \right).$$

$$(4.80)$$

However, comparing the coefficient of like power of p, we obtain

$$p^0 : u_0(x,t) = x^4 + ty^4$$

$$p^1 : u_1(x,t) = x^4 \frac{\left(\left(t + \frac{1}{\Gamma(\beta)}\right)^\beta - \left(\frac{1}{\Gamma(\beta)}\right)^\beta \right)^2}{2!(\beta)} + y^4 \frac{\left(\left(t + \frac{1}{\Gamma(\beta)}\right)^\beta - \left(\frac{1}{\Gamma(\beta)}\right)^\beta \right)^3}{3!(\beta)}$$

$$p^2 : u_2(x,t) = x^4 \frac{\left(\left(t + \frac{1}{\Gamma(\beta)}\right)^\beta - \left(\frac{1}{\Gamma(\beta)}\right)^\beta \right)^4}{4!(\beta)} + y^4 \frac{\left(\left(t + \frac{1}{\Gamma(\beta)}\right)^\beta - \left(\frac{1}{\Gamma(\beta)}\right)^\beta \right)^5}{5!(\beta)}$$

$$p^3 : u_3(x,t) = x^4 \frac{\left(\left(t + \frac{1}{\Gamma(\beta)}\right)^\beta - \left(\frac{1}{\Gamma(\beta)}\right)^\beta \right)^6}{6!(\beta)} + y^4 \frac{\left(\left(t + \frac{1}{\Gamma(\beta)}\right)^\beta - \left(\frac{1}{\Gamma(\beta)}\right)^\beta \right)^7}{7!(\beta)}$$

$$\vdots$$

$$p^n : u_n(x,y,t) = x^4 \frac{\left(\left(t + \frac{1}{\Gamma(\beta)}\right)^\beta - \left(\frac{1}{\Gamma(\beta)}\right)^\beta \right)^{2n}}{(2n)!(\beta)} + y^4 \frac{\left(\left(t + \frac{1}{\Gamma(\beta)}\right)^\beta - \left(\frac{1}{\Gamma(\beta)}\right)^\beta \right)^{2n+1}}{(2n+1)!(\beta)}.$$

$$(4.81)$$

Hence, the solution is given as

$$u(x,y,t) = \sum_{n}^{\infty} x^4 \frac{\left(\left(t + \frac{1}{\Gamma(\beta)}\right)^\beta - \left(\frac{1}{\Gamma(\beta)}\right)^\beta \right)^{2n}}{(2n)!(\beta)} + y^4 \frac{\left(\left(t + \frac{1}{\Gamma(\beta)}\right)^\beta - \left(\frac{1}{\Gamma(\beta)}\right)^\beta \right)^{2n+1}}{(2n+1)!(\beta)}$$

$$= x^4 \cosh_\beta(t) + y^4 \sinh_\beta(t). \qquad (4.82)$$

This is the exact solution of our equation.

Example 4.8. Consider the following three-dimensional inhomogeneous initial boundary value problem which describes the heat-like model:

$$u_t = x^4 y^4 z^4 + \frac{1}{36}(x^2 u_{xx} + x^2 u_{yy} + x^2 u_{zz}), \quad 0 < x, y, z < 1, \ t > 0. \quad (4.83)$$

Equation (4.83) is subjected to the following boundary conditions

$$u(0, y, z, t) = 0, \quad u(1, y, z, t) = y^4 z^4 (e^t - 1)$$

$$u(x, 0, z, t) = 0, \quad u(x, 1, z, t) = x^4 z^4 (e^t - 1)$$

$$u(x, y, 0, t) = 0, \quad u(x, y, 1, t) = y^4 x^4 (e^t - 1), \quad (4.84)$$

and initial condition

$$u(x, y, z, 0) = 0. \quad (4.85)$$

Following the step involved in the described technique, we obtain

$$\sum_{n=0}^{\infty} p^n u_n(x, y, t) = x^4 y^4 z^4 t + p\left(\frac{1}{36}x^2 S^{-1}\left[uS_\beta\left(\sum_{n=0}^{\infty} p^n u_n(x, y, t)\right)_{xx}\right]\right)$$

$$+ p\left(\frac{1}{36}y^2 S^{-1}\left[u^2 S_\beta\left(\sum_{n=0}^{\infty} p^n u_n(x, y, t)\right)_{yy}\right]\right)$$

$$+ p\left(\frac{1}{36}z^2 S^{-1}\left[u^2 S_\beta\left(\sum_{n=0}^{\infty} p^n u_n(x, y, t)\right)_{yy}\right]\right). \quad (4.86)$$

However, comparing the coefficient of like power of p, we obtain

$$p^0 : u_0(x, t) = x^4 y^4 z^4 t$$

$$p^1 : u_1(x, t) = x^4 y^4 z^4 \frac{\left(\left(t + \frac{1}{\Gamma(\beta)}\right)^\beta - \left(\frac{1}{\Gamma(\beta)}\right)^\beta\right)^2}{2!(\beta)}$$

$$p^2 : x^4 y^4 z^4 \frac{\left(\left(t + \frac{1}{\Gamma(\beta)}\right)^\beta - \left(\frac{1}{\Gamma(\beta)}\right)^\beta\right)^3}{3!(\beta)}$$

$$p^3 : u_3(x, t) = x^4 y^4 z^4 \frac{\left(\left(t + \frac{1}{\Gamma(\beta)}\right)^\beta - \left(\frac{1}{\Gamma(\beta)}\right)^{2-\beta}\right)^5}{5!(2 - \beta)}$$

$$\vdots$$

$$p^n : u_n(x, y, t) = x^4 y^4 z^4 \frac{\left(\left(t + \frac{1}{\Gamma(\beta)}\right)^\beta - \left(\frac{1}{\Gamma(\beta)}\right)^\beta\right)^{2n+1}}{(2n+1)!(\beta)}. \tag{4.87}$$

We obtain the solution as

$$\sum_{n=0}^{\infty} x^4 y^4 z^4 \frac{\left(\left(t + \frac{1}{\Gamma(\beta)}\right)^\beta - \left(\frac{1}{\Gamma(\beta)}\right)^\beta\right)^{2n+1}}{(2n+1)!(\beta)} = x^4 y^4 z^4 (\exp_\beta(t) - 1). \tag{4.88}$$

4.5 LAPLACE DECOMPOSITION METHOD

The Laplace transform method has a number of properties that make it useful for analyzing linear dynamical systems. The most significant advantage is that differentiation and integration become multiplication and division, respectively, by s similarly to logarithms changing multiplication of numbers to addition of their logarithms. Because of this property, the Laplace variable s is also known as an operator variable in the L domain: either derivative operator or for s^1 integration operator. The transform method turns integral equations and differential equations into polynomial equations, which are much easier to solve. Once solved, use of the inverse Laplace transform method reverts to the time domain. However, when dealing with nonlinear equations, the Laplace transform method can not be used alone. An additional technique combined with the Laplace transform method can be used to derive the approximate or exact solution of the equation depending on the complexity of the nonlinear part. In particular, linear and nonlinear partial differential equations with beta-derivative cannot be solved by Laplace transform method. We therefore used the beta-derivative presented earlier. We consider a general nonlinear nonhomogenous partial differential equation with the initial conditions of the form:

$$_0^A D_t^\beta (u(x, t)) + L(u(x, t)) + N(u(x, t)) = g(x, t), \quad u(x, 0) = h(x), \quad u_t(x, 0) = f(x). \tag{4.89}$$

L is the linear differential operator, N represents the general nonlinear differential operator, and $g(x, t)$ is the source term. Taking the beta-Sumudu transform operator on both sides of Equation (4.89), we get

$$L_\beta\left({}_0^A D_t^\beta(u(x,t))\right) + L_\beta(L(u(x,t))) + L_\beta(N(u(x,t))) = L_\beta(g(x,t))(u).$$
(4.90)

Using the differentiation property of the Laplace transform method and above initial conditions, we have

$$L_\beta(u(x,t))(s) = \frac{1}{s}L_\beta(g(x,t)) + \frac{1}{s}f(x) - \frac{1}{s}L_\beta(L(u(x,t))) - \frac{1}{s}L_\beta(N(u(x,t))).$$
(4.91)

Now, applying the well-known inverse Laplace transform method on both sides of Equation (4.91), we get:

$$u(x,t) = F(x,t) - L^{-1}\left[-\frac{1}{s}L_\beta(L(u(x,t))) - \frac{1}{s}L_\beta(N(u(x,t)))\right], \quad (4.92)$$

where $F(x,t)$ represents the term arising from the source term and the prescribed initial conditions. Now, we apply the homotopy perturbation method

$$u(x,t) = \lim_{p\to 1}\sum_{n=0}^{\infty} p^n u_n(x,t), \tag{4.93}$$

and the nonlinear term can be decomposed as

$$N(u(x,t)) = \sum_{n=0}^{\infty} p^n H_n(u), \tag{4.94}$$

where

$$H_n(u_0,u_1,\ldots,u_n) = \frac{1}{n!}\frac{\partial^n}{\partial p^n}\left[N\left(\sum_{i=0}^{\infty} p^i u(x,t)\right)\right]. \tag{4.95}$$

Replacing this into Equation (4.89)

$$\sum_{n=0}^{\infty} p^n u_n(x,t) = F(x,t) - pL^{-1}\left[-\frac{1}{s}L_\beta\left(L\left(\sum_{n=0}^{\infty} p^n u_n(x,t)\right)\right)\right.$$
$$\left. -\frac{1}{s}L_\beta\left(\sum_{n=0}^{\infty} p^n H_n(u)\right)\right]. \tag{4.96}$$

After a comparison of coefficients of like power p, the following components are obtained

$$p^0 : u_0(x,t) = F(x,t)$$

$$p^1 : u_1(x,t) = -L^{-1}\left[\frac{1}{s}L_\beta[Lu_0(x,t) + H_0(u_0)]\right]$$

$$p^2 : u_2(x,t) = -L^{-1}\left[\frac{1}{s}S_\beta[Lu_1(x,t) + H_0(u_0,u_1)]\right]$$

$$p^3 : u_3(x,t) = -L^{-1}\left[\frac{1}{s}L_\beta[Lu_2(x,t) + H_0(u_0,u_1,u_2)]\right]$$

$$\vdots$$

$$p^n : u_n(x,t) = -L^{-1}\left[\frac{1}{s}L_\beta[Lu_{(n-1)}(x,t) + H_0(u_0,u_1,u_2,\ldots,u_{(n-1)})]\right].$$

$$(4.97)$$

We illustrate the above technique by solving some partial beta-equations.

Example 4.9. Consider the following two-dimensional initial boundary value problem describing the model of wave-like

$$u_{tt} = \frac{1}{12}(x^2 u_{xx} + y^2 u_{yy}), \quad 0 < x,y < 1, \quad t > 0, \tag{4.98}$$

subject to the Neumann boundary conditions

$$u_x(0,y,t) = u_y(x,0,t) = 0, \quad u_x(1,y,t) = 4\cosh(t), \quad u_x(x,1,t) = 4\sinh(t), \tag{4.99}$$

and initial conditions: $u(x,y,0) = x^4, u_t(x,y,0) = y^4$.

Using the routine presented in this section, we have

$$\sum_{n=0}^{\infty} p^n u_n(x,y,t) = x^4 + y^4 t + p\left(\frac{1}{12}x^2 L^{-1}\left[\frac{1}{s^2}L_\beta\left(\sum_{n=0}^{\infty} p^n u_n(x,y,t)\right)_{xx}\right]\right)$$

$$+ p\left(\frac{1}{12}y^2 L^{-1}\left[\frac{1}{s^2}L_\beta\left(\sum_{n=0}^{\infty} p^n u_n(x,y,t)\right)_{yy}\right]\right).$$

$$(4.100)$$

However, comparing the coefficient of like power of p, we obtain

$$p^0 : u_0(x,t) = x^4 + ty^4$$

$$p^1 : u_1(x,t) = x^4\frac{\left(\left(t+\frac{1}{\Gamma(\beta)}\right)^\beta - \left(\frac{1}{\Gamma(\beta)}\right)^\beta\right)^2}{2!(\beta)} + y^4\frac{\left(\left(t+\frac{1}{\Gamma(\beta)}\right)^\beta - \left(\frac{1}{\Gamma(\beta)}\right)^\beta\right)^3}{3!(\beta)}$$

$$p^2 : u_2(x,t) = x^4\frac{\left(\left(t+\frac{1}{\Gamma(\beta)}\right)^\beta - \left(\frac{1}{\Gamma(\beta)}\right)^\beta\right)^4}{4!(\beta)} + y^4\frac{\left(\left(t+\frac{1}{\Gamma(\beta)}\right)^\beta - \left(\frac{1}{\Gamma(\beta)}\right)^\beta\right)^5}{5!(\beta)}$$

$$p^3 : u_3(x,t) = x^4 \frac{\left(\left(t + \frac{1}{\Gamma(\beta)}\right)^\beta - \left(\frac{1}{\Gamma(\beta)}\right)^\beta\right)^6}{6!(\beta)} + y^4 \frac{\left(\left(t + \frac{1}{\Gamma(\beta)}\right)^\beta - \left(\frac{1}{\Gamma(\beta)}\right)^\beta\right)^7}{7!(\beta)}$$

$$\vdots$$

$$p^n : u_n(x,y,t) = x^4 \frac{\left(\left(t + \frac{1}{\Gamma(\beta)}\right)^\beta - \left(\frac{1}{\Gamma(\beta)}\right)^\beta\right)^{2n}}{(2n)!(\beta)} + y^4 \frac{\left(\left(t + \frac{1}{\Gamma(\beta)}\right)^\beta - \left(\frac{1}{\Gamma(\beta)}\right)^\beta\right)^{2n+1}}{(2n+1)!(\beta)}.$$

(4.101)

Hence, the solution is given as

$$u(x,y,t) = \sum_n^\infty x^4 \frac{\left(\left(t + \frac{1}{\Gamma(\beta)}\right)^\beta - \left(\frac{1}{\Gamma(\beta)}\right)^\beta\right)^{2n}}{(2n)!(\beta)}$$
$$+ y^4 \frac{\left(\left(t + \frac{1}{\Gamma(\beta)}\right)^\beta - \left(\frac{1}{\Gamma(\beta)}\right)^\beta\right)^{2n+1}}{(2n+1)!(\beta)}$$
$$= x^4 \cosh_\beta(t) + y^4 \sinh_\beta(t).$$

(4.102)

This is the exact solution of our equation.

Example 4.10. Consider the following three-dimensional inhomogeneous initial boundary value problem which describes the heat-like model:

$$u_t = x^4 y^4 z^4 + \frac{1}{36}(x^2 u_{xx} + x^2 u_{yy} + x^2 u_{zz}), \quad 0 < x,y,z < 1, \quad t > 0.$$

(4.103)

Equation (4.103) is subjected to the following boundary conditions

$$u(0,y,z,t) = 0, \quad u(1,y,z,t) = y^4 z^4 (e^t - 1)$$
$$u(x,0,z,t) = 0, \quad u(x,1,z,t) = x^4 z^4 (e^t - 1)$$
$$u(x,y,0,t) = 0, \quad u(x,y,1,t) = y^4 x^4 (e^t - 1),$$

(4.104)

and initial condition

$$u(x,y,z,0) = 0.$$

(4.105)

Following the step involved in the described technique, we obtain

$$\sum_{n=0}^\infty p^n u_n(x,y,t) = x^4 y^4 z^4 t + p\left(\frac{1}{36}x^2 L^{-1}\left[\frac{1}{s}L_\beta\left(\sum_{n=0}^\infty p^n u_n(x,y,t)\right)_{xx}\right]\right)$$

$$+ p \left(\frac{1}{36} y^2 L^{-1} \left[\frac{1}{s^2} L_\beta \left(\sum_{n=0}^{\infty} p^n u_n(x, y, t) \right)_{yy} \right] \right)$$

$$+ p \left(\frac{1}{36} z^2 L^{-1} \left[\frac{1}{s^2} L_\beta \left(\sum_{n=0}^{\infty} p^n u_n(x, y, t) \right)_{yy} \right] \right).$$

$$(4.106)$$

However, comparing the coefficient of like power of p, we obtain

$$p^0 : u_0(x, t) = x^4 y^4 z^4 t$$

$$p^1 : u_1(x, t) = x^4 y^4 z^4 \frac{\left(\left(t + \frac{1}{\Gamma(\beta)} \right)^\beta - \left(\frac{1}{\Gamma(\beta)} \right)^\beta \right)^2}{2!(\beta)}$$

$$p^2 : x^4 y^4 z^4 \frac{\left(\left(t + \frac{1}{\Gamma(\beta)} \right)^\beta - \left(\frac{1}{\Gamma(\beta)} \right)^\beta \right)^3}{3!(\beta)}$$

$$p^3 : u_3(x, t) = x^4 y^4 z^4 \frac{\left(\left(t + \frac{1}{\Gamma(\beta)} \right)^\beta - \left(\frac{1}{\Gamma(\beta)} \right)^{2-\beta} \right)^5}{5!(\beta)}$$

$$\vdots$$

$$p^n : u_n(x, y, t) = x^4 y^4 z^4 \frac{\left(\left(t + \frac{1}{\Gamma(\beta)} \right)^\beta - \left(\frac{1}{\Gamma(\beta)} \right)^\beta \right)^{2n+1}}{(2n+1)!(\beta)}. \qquad (4.107)$$

We obtain the solution as

$$\sum_{n=0}^{\infty} x^4 y^4 z^4 \frac{\left(\left(t + \frac{1}{\Gamma(\beta)} \right)^\beta - \left(\frac{1}{\Gamma(\beta)} \right)^\beta \right)^{2n+1}}{(2n+1)!(\beta)} = x^4 y^4 z^4 (\exp_\beta(t) - 1).$$

$$(4.108)$$

4.6 EXTENSION OF MATCH ASYMPTOTIC METHOD TO FRACTIONAL BOUNDARY LAYERS PROBLEMS

In mathematics, the method of matched asymptotic expansions is a common approach to finding an accurate approximation to the solution to an equation, or system of equations. It is particularly used when solving singularly perturbed differential equations. It involves finding several different approximate solutions, each of which is valid that is accurate for part of the range of the independent variable, and then combining these different solutions together to give a single approximate solution that is valid for the whole range of values of the independent variable. In a large class of singularly perturbed problems, the domain may be divided into two or more subdomains. In one of these, often the largest, the solution is accurately approximated by an asymptotic series found by treating the problem as a regular perturbation; that is, by setting a relatively small parameter to zero. The other subdomains consist of one or more small areas in which that approximation is inaccurate, generally because the perturbation terms in the problem are not negligible there. These areas are referred to as transition layers, and as boundary or interior layers depending on whether they occur at the domain boundary as is the usual case in applications or inside the domain. However, the method of match asymptotic cannot be used in the case of fractional ordinary and partial differential equations. In order to further extend the method of match asymptotic method to the concept of fractional order derivative, we model boundary layers problems with beta-derivative.

4.6.1 Methodology
Consider the following fractional singular perturbation problem

$$P^\beta(u(x,\epsilon)), \quad u(a) = A, \quad u(b) = B, \tag{4.109}$$

where $\beta \geq 1$ is the fractional order. To solve Equation (4.109), we assume that

$$u(x,\epsilon) = \sum_{n=0}^{\infty} \epsilon^n u_n(x). \tag{4.110}$$

We can then find the outer solution using the same routine as in the case of ordinary differential equation with integer order derivatives. To find the inner solution, we also follow the same routine as in the case of differential equation with integer order derivative, with the difference that after changing the variable we make use of Atangana beta-rule provided in Chapter 2. We shall illustrate this with an example.

Example 4.11. Consider the following beta-singular perturbation problem

$$\epsilon u'' + (1+\epsilon)_0^A D_x^\beta y + y = 0, \quad 0 < \epsilon \ll 1, \tag{4.111}$$

subjected to the following initial and boundary conditions

$$u(0) = 0, \quad u(1) = 1. \tag{4.112}$$

The outer solution is valid for $x = O(1)$; this will imply that Equation (4.111) is reduced to

$$_0^A D_x^\beta y + y = 0, \tag{4.113}$$

for which the exact solution is provided

$$u(x, \beta) = c \exp\left[-\frac{\left(x + \frac{1}{\Gamma(\beta)}\right)^\beta}{\beta} \right]. \tag{4.114}$$

If we apply directly the initial condition, we obtain that $c = 0$. But applying the boundary condition, we obtain $c = \exp\left[\frac{\left(1 + \frac{1}{\Gamma(\beta)}\right)^\beta}{\beta} \right]$, which is not possible as c has two values. Therefore, according the match asymptotic technique, we consider the outer solution where $c = \exp\left[\frac{\left(1 + \frac{1}{\Gamma(\beta)}\right)^\beta}{\beta} \right]$, therefore this last one leads to the solution of the outer solution in the outer region. The inner solution is valid only for $t = O(\epsilon)$. In the inner region, t and ϵ are both very small, but of comparable size, so define the new $O(\zeta)$ time variable $\zeta = \frac{x}{\epsilon}$. Rescaling the original fractional boundary layer problem by replacing $x = \zeta\epsilon$, the problem becomes

$$\frac{1}{\epsilon} u''(\zeta) + (1 + \epsilon)_0^A D_x^\beta y(\zeta) + y(\zeta) = 0. \tag{4.115}$$

Of course, by multiplying by the small parameter and taking it to be zero, we shall obtain

$$u''(\zeta) + _0^A D_x^\beta y(\zeta) = 0. \tag{4.116}$$

The exact solution of this equation is given as

$$u(\zeta, \beta) = C \int \exp\left[-\frac{\left(\zeta + \frac{1}{\Gamma(\beta)}\right)^\beta}{\beta} \right] d\zeta$$

$$= C\left(\left(\zeta + \frac{1}{\Gamma(\beta)}\right)\right) \text{ExpIntegralE}\left(1 + \frac{1}{\beta}, \frac{\zeta + \left(\frac{1}{\Gamma(\beta)}\right)^\beta}{\beta}\right) \beta^{-1} + B, \tag{4.117}$$

where ExpIntegralE is the exponential integral function defined as

$$\text{ExpIntegralE}\,[n, x] = \int_x^\infty \frac{\exp[-xt]}{t^n}. \tag{4.118}$$

To find the constants B and C, we apply the boundary conditions. First, for $t = 0$, which corresponds to the inner region, we have

$$u(\zeta,\beta) = C\left(\frac{1}{\Gamma(\beta)}\right) \text{ExpIntegralE}\left(1 + \frac{1}{\beta}, \frac{\left(\frac{1}{\Gamma(\beta)}\right)^{\beta}}{\beta}\right)\beta^{-1} + B = 0,$$

(4.119)

such that the outer solution becomes

$$u(\zeta,\beta) = -C\left(\frac{1}{\Gamma(\beta)}\right) \text{ExpIntegralE}\left(1 + \frac{1}{\beta}, \frac{\left(\frac{1}{\Gamma(\beta)}\right)^{\beta}}{\beta}\right)\beta^{-1}$$

$$+ C\left(\zeta + \frac{1}{\Gamma(\beta)}\right) \text{ExpIntegralE}\left(1 + \frac{1}{\beta}, \frac{\zeta + \left(\frac{1}{\Gamma(\beta)}\right)^{\beta}}{\beta}\right)\beta^{-1}.$$

(4.120)

Currently working to find the constant C, we employ the harmonizing method also called the matching method. The principal idea here is that the inner and the outer solutions should harmonize for values of x in an intermediary region, where $\epsilon \ll t \ll 1$ demands the outer limit of the inner solution to match the limit of the outer solution, which in mathematical formula implies

$$\lim_{\zeta \to \infty} u_I = \lim_{x \to 0} u_O.$$

(4.121)

This corresponds to

$$C^{-1} = -\left(\frac{1}{\Gamma(\beta)}\right) \text{ExpIntegralE}\left(1 + \frac{1}{\beta}, \frac{\left(\frac{1}{\Gamma(\beta)}\right)^{\beta}}{\beta}\right)\beta^{-1}.$$

(4.122)

Composite solution. To have the final matched or composite solution, which is valid, on the whole domain, one of the commonly used methods is the so-called uniform method. This method consists of an addition of the inner and the outer approximations and subtracting their overlapping value u_{overlap}, which would otherwise be counted twice. The overlapping value is the limit of the inner boundary layer solution and the inner limit of the outer solution. In mathematical formula, we have the following:

$$u(x,\epsilon) = u_I(x,\epsilon) + u_O(x,\epsilon) - u_{\text{overlap}}.$$

(4.123)

4.7 NUMERICAL METHOD

Solving difficult equations with numerical scheme has been a passionate exercise for many scholars [101–103]. However, there exist numerous versions of this scheme in the literature [101–103]. Some of these numerical techniques are very accurate while approximating solutions of difficult equations. These numerical methods yield approximate solutions to the governing equation through the discretization of space and time. Within the discredited problem domain, the variable internal properties, boundaries, and stresses of the system are approximated. Deterministic, distributed-parameter, numerical models can relax the rigid idealized conditions of analytical models or lumped-parameter models, and they can therefore be more realistic and flexible for simulating field conditions. In this section, we present an application of some numerical technique for some beta partial differential equations. Let us consider the following convection-diffusion equation.

Example 4.12.

$$
{}_0^A D_t^\beta T(x,t) = a\frac{\partial T(x,t)}{\partial x^2} - \epsilon u\frac{\partial T(x,t)}{\partial x} + \frac{Q}{c\rho}, \quad a = \frac{\lambda}{c\rho}, \qquad (4.124)
$$

with a the diffusion coefficient. A solution of beta-transient convection-diffusion equation can be approximated through the finite difference approach, known as finite difference method. We shall start with the explicit scheme. Note that the explicit scheme of finite difference method has been considered and stability criteria are formulated. In this scheme, temperature is totally dependent on the old temperature, meaning the initial conditions. Substitution of $\theta = 0$ gives the explicit discretization of the unsteady conductive heat transfer equation where θ is the weighing parameter between 0 and 1. We first recall that

$$
{}_0^A D_t^\beta \left(f(t_i)\right) = \left(t_i + \frac{1}{\Gamma(\beta)}\right)^{1-\beta}\frac{T_i^j - T_i^{j-1}}{\Delta t}. \qquad (4.125)
$$

Replacing this in Equation (4.124) and using the well-known approximation of second derivative, we obtain the following

$$
\left(t_i + \frac{1}{\Gamma(\beta)}\right)^{1-\beta}\frac{T_i^j - T_i^{j-1}}{\Delta t} = a\frac{T_{i-1}^{j-1} - 2T_i^{j-1} + T_{i+1}^{j-1}}{h^2}
$$

$$
- \epsilon u\frac{T_{i+1}^{j-1} - T_{i-1}^{j-1}}{\Delta t} + \frac{Q_i^{j-1}}{c\rho}, \qquad (4.126)
$$

where $\Delta = t^j - t^{j-1}$ and h is the uniform grid spacing or mesh step. Now dividing Equation (4.126) by $\left(t_i + \frac{1}{\Gamma(\beta)}\right)^{1-\beta}$ and simplicity, we put $a_1 = a\left(t_i + \frac{1}{\Gamma(\beta)}\right)^{\beta-1}$ and $\epsilon_1 = \epsilon\left(t_i + \frac{1}{\Gamma(\beta)}\right)^{\beta-1}$. Rearranging, we obtain the following recursive formula

$$T_i^j = \left(1 - \frac{2a_1\Delta t}{h^2}\right)T_i^{j-1} + \left(\frac{a_1\Delta t}{h^2} + \frac{\epsilon_1 u \Delta t}{2h}\right)T_{i-1}^{j-1}$$

$$+ \left(\frac{a_1\Delta t}{h^2} - \frac{\epsilon_1 u \Delta t}{2h}\right)T_{i+1}^{j-1} + \frac{Q_{i1}^{j-1}}{c\rho}. \qquad (4.127)$$

The criteria of the stability analysis reads

$$h < \frac{1}{\epsilon_1 u}, \quad \Delta t < \frac{h^2}{2a_1}. \qquad (4.128)$$

This inequality sets a stringent maximum limit to the time step size and represents a serious limitation for the explicit scheme. This method is not recommended for general transient problems, because the maximum possible time step has to be reduced as the square of h.

Example 4.13. Consider the normalized heat equation in one dimension, with homogeneous Dirichlet boundary conditions

$$_0^A D_t^u = u_{xx}, \quad u(0,t) = u(1,t) = 0, \quad u(x,0) = u_0(x). \qquad (4.129)$$

One way to solve this equation numerically is to approximate all the derivatives by finite differences. We partition the domain in space using a mesh $x_0, \ldots, x_M$ and in time using a mesh $t_0, \ldots, t_N$. We assume a uniform partition both in space and in time, so the difference between two consecutive space points will be h and between two consecutive time points will be k. The points

$$u(x_j, t_n) = u_j^n \qquad (4.130)$$

will represent the numerical approximation of $u(x_j, t_n) = u_j^n$. We present first the solution with explicit method. Using a forward difference at time t_n and a second-order central difference for the space derivative at position FTCS (Forward Time Centered Space), we get the recurrence equation:

$$\left(t_n + \frac{1}{\Gamma(\beta)}\right)^{1-\beta} \frac{u_i^{j+1} - u_i^j}{\Delta t} = \frac{u_{j+1}^n - 2u_j^n + u_{j-1}^n}{h^2}. \qquad (4.131)$$

This is an explicit method for solving the one-dimensional heat equation. We can obtain u_j^{n+1} from the other values this way:

$$u_j^{n+1} = (1+2k)u_i^j + ku_{j-1}^n + ku_{j+1}^n, \quad k = \max_n \left(\frac{\Delta t}{h^2 \left(t_n + \frac{1}{\Gamma(\beta)} \right)^{1-\beta}} \right).$$

$$(4.132)$$

So, with this recurrence relation, and knowing the values at time n, one can obtain the corresponding values at time $n + 1$. u_0^n and u_N^n must be replaced by the boundary conditions; in this example, they are both 0. The stability condition reads

$$\frac{\Delta t}{h^2} \le \max_n \left(t_n + \frac{1}{\Gamma(\beta)} \right)^{1-\beta}.$$

$$(4.133)$$

The numerical errors are proportional to the time step and the square of the space step:

$$\Delta u = O(\Delta t) + O(h^2).$$

$$(4.134)$$

Implicit scheme method. If we use the backward difference at time t_{n+1} and a second-order central difference for the space derivative at position x_j (The Backward Time, Centered Space Method) we get the recurrence equation:

$$\frac{u_j^{n+1} - u_j^n}{\Delta t} = \frac{u_{j+1}^{n+1} - 2u_j^{n+1} + u_{j-1}^{n+1}}{h^2}.$$

$$(4.135)$$

This is an implicit method for solving the one-dimensional heat equation.

We can obtain u_j^{n+1} from solving a system of linear equations:

$$(1 + 2\Delta t)u_j^{n+1} - ku_{j-1}^{n+1} - ku_{j+1}^{n+1} = u_j^n.$$

$$(4.136)$$

The scheme is always numerically stable and convergent, but usually more numerically intensive than the explicit method, as it requires solving a system of numerical equations on each time step. The errors are linear over the time step and quadratic over the space step:

$$\Delta u = O(\Delta t) + O(h^2).$$

$$(4.137)$$

Crank–Nicolson method. Finally if we use the central difference at time $t_{n+1/2}$ and a second-order central difference for the space derivative at position x_j, we get the recurrence equation:

$$\frac{u_j^{n+1} - u_j^n}{k} = \frac{1}{2}\left(\frac{u_{j+1}^{n+1} - 2u_j^{n+1} + u_{j-1}^{n+1}}{h^2} + \frac{u_{j+1}^n - 2u_j^n + u_{j-1}^n}{h^2}\right). \quad (4.138)$$

This formula is known as the Crank–Nicolson method.

We can obtain u_j^{n+1} from solving a system of linear equations:

$$(2 + 2k)u_j^{n+1} - ku_{j-1}^{n+1} - ku_{j+1}^{n+1} = (2 - 2k)u_j^n + ku_{j-1}^n + ku_{j+1}^n. \quad (4.139)$$

The scheme is always numerically stable and convergent, but usually more numerically intensive, as it requires solving a system of numerical equations on each time step. The errors are quadratic over both the time step and the space step:

$$\Delta u = O(\Delta t^2) + O(h^2). \quad (4.140)$$

Usually the Crank–Nicolson scheme is the most accurate scheme for small time steps. The explicit scheme is the least accurate and can be unstable, but is also the easiest to implement and the least numerically intensive. The implicit scheme works the best for large time steps.

4.8 GENERALIZED STATIONARITY WITH A NEW PARAMETER

4.8.1 Generalized time evolution

Mathematical models are usually formulated as initial value problems for dynamical evolution equations written as [104]:

$$\frac{d}{dt}p(t) = Ap(t), \quad (4.141)$$

where t is the time taken from $\mathbb{R}_+$ and A is an operator in a Banach space. The aim here is to find the state $p(t)$ of the model at a time $t > t_0$ depending on the initial state $p(t_0)$ at the initial time t_0. Many scientists around the world have tried to extend classical models to models with fractional derivative (see [75, 78, 79, 90]) and analyze them with various methods in order to provide a broader view on the natural phenomena under investigation. For example, the authors [78, 104] successfully extended the advection-dispersion equation (to the fractional one) by using various techniques including the well-known action of the Fourier transform method on integer derivatives to rational order. However, generalizing the model (4.141), by substituting the time differentiation $\frac{d}{dt}$ with a derivative $\frac{d^\sigma}{dt^\sigma}$ of fractional order $\sigma > 0$ to obtain the following model

$$\frac{d^{\sigma}}{dt^{\sigma}} p(t) = Ap(t) \qquad (4.142)$$

has raised a number of fundamental questions [48–50] and is still dividing the scientific community. The term $\frac{d}{dt}$ is seen as the representation of the rate of accumulation or loss in the system and mainly reflects the basic principle of locality, together with the time translation stationarity. Moreover, we know from the classical calculus that

$$\frac{d}{dx} g(x) = \lim_{t \to 0} \frac{g(x) - g(x - t)}{t} = - \lim_{t \to 0} \frac{g(t)g(x) - g(x)}{t}. \qquad (4.143)$$

This means that $-\frac{d}{dt}$ defines the infinitesimal generator of the time translations given by

$$g(t)g(x) = g(x - t).$$

In a similar manner, fractional derivative D_t^{γ} of order $0 < \gamma < 1$ can be defined [104–107] as

$$D_t^{\gamma}(g(\tau)) = - \lim_{t \to 0} \frac{g_{\gamma}(t)g(\tau) - g(\tau)}{t}, \qquad (4.144)$$

where $g_{\gamma}(t)$ the fractional time evolution, considered as universal attractor semigroups of coarse-grained macroscopic time evolutions. For instance, it is shown that [75, 76, 78, 104]:

$$D_t^{\gamma}(g(\tau)) = -\frac{1}{\Gamma(-\gamma)} \int_0^{\infty} \frac{g(\tau - r) - g(\tau)}{r^{\gamma + 1}} \, dr, \quad 0 < \gamma < 1, \quad (4.145)$$

which is the fractional derivative of $g(t)$ in the sense of Marchaud.

Definition 4.8.1. A time evolution is a pair $(\{\mathfrak{T}_{\beta}(t), 0 \leq t < \infty\}, (X_{\beta}, \| \cdot \|))$ with $\mathfrak{T}_{\beta}(t) = \mathfrak{T}(t\beta)$ defining a semigroup of operators $(\{\mathfrak{T}_{\beta}(t), 0 \leq t < \infty\}$ mapping the Banach space $(X_{\beta}(\mathbb{R}), \| \cdot \|)$ of functions $g_{\beta}(x) = g(x\beta)$ on $\mathbb{R}$ to itself.

In the expression $\mathfrak{T}_{\beta}(t)$, the variable $t > 0$ represents a time duration and the variable $x \in \mathbb{R}$ in the expression $g_{\beta}(x)$ stands for a time instant. The index $\beta > 0$ indicates the units of time. The elements $g_{\beta}(x) = g(x\beta)$, as functions of the time coordinates $x \in \mathbb{R}$, represent observable states of a given physical system.

4.8.2 Basic settings for time evolutions [108]

- Semigroup: The following conditions define the semigroup:
$$\mathfrak{T}_{\beta}(t_1)\mathfrak{T}_{\beta}(t_2)g_{\beta}(t_0) = \mathfrak{T}_{\beta}(t_1 + t_2)g_{\beta}(t_0)$$
$$\mathfrak{T}_{\beta}(0)g_{\beta}(t_0) = g_{\beta}(t_0),$$

with $t_1, t_2 > 0, t_0 \in \mathbb{R}$ and $g_\beta \in X_\beta$.

- Homogeneity of the time argument t: This requires the commutativity with translations
$$\mathfrak{T}_\beta(t_1)\mathcal{T}_\beta(t_2)g_\beta(t_0) = \mathcal{T}_\beta(t_2)\mathfrak{T}_\beta(t_1)g_\beta(t_0),$$

with $t_2 > 0, t_1, t_0 \in \mathbb{R}$. Hence, this allows to shift the origin of time and it reflects the basic symmetry of time translation invariance.

- Continuity: We assume that the time evolution is strongly continuous in t such that
$$\lim_{t \longrightarrow 0} \|\mathfrak{T}_\beta(t)g_\beta - g_\beta\| = 0$$

for all $g_\beta \in X_\beta$.

- Causality: Operator of the time evolution should be causal so that the function $g_\beta(t_0) = (\mathfrak{T}_\beta(t)f_\beta)(t_0)$ only depends on the values of $f_\beta(x)$ for $x < t_0$.

- Coarse-graining: The time evolution operator $\mathfrak{T}_\beta(t)$ should be establishable using the procedure of a coarse-graining. The main idea here is to combine a time average $\frac{1}{t}\int_{x-t}^{x} f_\beta(\xi)\mathrm{d}\xi$ when $t, x \longrightarrow \infty$ with a rescaling of x and t.

4.8.3 Diffusion using derivative with a new parameter

$$\begin{cases} {}^{A}_{0}D^{\beta}_{t}g(t,x) = \Delta g(t,x), & 0 < \beta < 1, \ t > 0, \ x > 0 \\ g(0,x) = f(x), & x > 0. \end{cases} \qquad (4.146)$$

To show the existence result for this model, we use the separation of variables technique and set $g(t,x) = T(t)X(x)$. Substitution in Equation (4.146) gives

$$X(x){}^{A}_{0}D^{\beta}_{t}T(t) = T(t)\Delta X(x)$$

or

$$\frac{{}^{A}_{0}D^{\beta}_{t}T(t)}{T(t)} = \frac{\Delta X(x)}{X(x)}.$$

We put $-\lambda = \frac{{}^{A}_{0}D^{\beta}_{t}T(t)}{T(t)} = \frac{\Delta X(x)}{X(x)}$ to get the Eigen-value system

$$\Delta X(x) = -\lambda X(x), \quad x > 0, \tag{4.147}$$

$$_0^A D_t^\beta T(t) = -\lambda T(t), \quad t > 0. \tag{4.148}$$

To solve the Eigen-value system (4.148), we use an infinite sequence of pairs $\{\alpha_n, \delta_n\}_{n \in \mathbb{N}}$ with $\{\alpha_n\}$ an increasing sequence such that $\alpha_n \to \infty$ and $\{\delta_n\}$ a family of functions that form a complete orthogonal set in $L^2((x_0, \infty))$. Exploiting α_n defined from Equation (4.148), we can find a solution of the Eigen-value problem for the β-derivative Equation (4.148) by putting $\lambda = \alpha_n$ (see [109]). Making use of Equation (4.148), the expression:

$$\mathcal{E}_\beta(t) = \exp\left(-\mu\left(\frac{\left(t + \frac{1}{\Gamma(\beta)}\right)^\beta - \Gamma(\beta)^{-\beta}}{\beta}\right)\right) \tag{4.149}$$

is the unique solution of the Eigen-value problem

$$_0^A D_t^\beta T(t) = -\mu T(t), \quad t > 0 \ \ T(0) = 1. \tag{4.150}$$

Therefore, the solution to Equation (4.148) is given as

$$T(t) = \tilde{f}(n) \exp\left(-\lambda\left(\frac{\left(t + \frac{1}{\Gamma(\beta)}\right)^\beta - \Gamma(\beta)^\beta}{\beta}\right)\right), \tag{4.151}$$

where $\tilde{f}(n)$ is chosen to satisfy the initial condition f. This leads us to a formal solution of the fractional Cauchy problem given by

$$g(t, x) = \sum_{n=1}^{\infty} \tilde{f}(n) \mathcal{E}_\beta(t) \delta_n(x). \tag{4.152}$$

Making use of relatively new concepts not yet in the literature, like the modified version of the Sumudu transform operator, we have analyzed some type of stationary states using the beta-derivative, a time derivative with a new parameter. We have proven that this new time derivative recovers the classical well-known stationarity results. However, we have shown that, contrary to most of the existing versions of fractional derivative, the time diffusion and relaxation systems, using the beta-derivative, are not governed by the Mittag–Leffler functions, but rather by a relatively newly introduced beta-exponential function $\mathcal{E}$. This is the first instance where such a result has been obtained and will lead to more investigations and results about the time derivative with a new parameter [104].

CHAPTER 5

Applications of local derivative with new parameter

5.1 INTRODUCTION

Derivatives are constantly used in everyday life to help measure how much something is changing. Derivative is used in every branch of the physical sciences, actuarial science, computer science, statistics, engineering, economics, business, medicine, demography, and in other fields wherever a problem can be mathematically modeled and an optimal solution is desired. It allows one to go from nonconstant rates of change to the total, change or vice versa, and many times in studying a problem we know one and are trying to find the other. Physics makes particular use of derivative; all concepts in classical mechanics and electromagnetism are related through calculus. The mass of an object of known density, the moment of inertia of objects, as well as the total energy of an object within a conservative field can be found by the use of calculus. An example of the use of calculus in mechanics is Newton's second law of motion: historically stated, it expressly uses the term *rate of change* which refers to the derivative, saying that the rate of change of momentum of a body is equal to the resultant force acting on the body and is in the same direction. Commonly expressed today as Force = Mass × Acceleration, it involves differential calculus because acceleration is the time derivative of velocity or second time derivative of trajectory or spatial position. Starting from knowing how an object is accelerating, we use calculus to derive its path. Maxwell's theory of electromagnetism and Einstein's theory of general relativity are also expressed in the language of differential calculus. Chemistry also uses calculus in determining reaction rates and radioactive decay. In biology, population dynamics starts with reproduction and death rates to model population changes.

Derivative with a New Parameter. http://dx.doi.org/10.1016/B978-0-08-100644-3.00005-2
© 2016 Elsevier Ltd. All rights reserved.

5.2 MODEL OF GROUNDWATER FLOW WITHIN THE CONFINED AQUIFER

An aquifer is an underground layer of water-bearing permeable rock or unconsolidated materials (gravel, sand, or silt) from which groundwater can be extracted using water well. The study of water flow in aquifers and the characterization of aquifers is called hydrogeology. Related terms include aquitard, which is a bed of low permeability along an aquifer and aquiclude or aquifuge, which is a solid, impermeable area underlying or overlying an aquifer. If the impermeable area overlies the aquifer, pressure could cause it to become a confined aquifer. Used in hydrogeology, the groundwater flow equation is the mathematical relationship which is used to describe the flow of groundwater through an aquifer. The transient flow of groundwater is described by a form of the diffusion equation, similar to that used in heat transfer to describe the flow of heat in a solid heat conduction. The steady-state flow of groundwater is described by a form of the Laplace equation, which is a form of potential flow and has analogs in numerous fields. The groundwater flow equation is often derived for a small representative elemental volume, where the properties of the medium are assumed to be effectively constant. A mass balance is done on the water flowing in and out of this small volume, the flux terms in the relationship being expressed in terms of head by using the constitutive equation called Darcy's law, which requires that the flow is slow. Most commonly, an aquifer test is conducted by pumping water from one well at a steady rate and for at least 1 day, while carefully measuring the water levels in the monitoring wells. When water is pumped from the pumping well, the pressure in the aquifer that feeds that well declines. This decline in pressure will show up as drawdown change in hydraulic head in an observation well. Drawdown decreases with radial distance from the pumping well and drawdown increases with the length of time that the pumping continues. The following figure shows the cross-sectional view of a simple confined or unconfined aquifer. The aquifer characteristics that are evaluated by most aquifer tests are as follows (Figure 5.1):

1. *Hydraulic conductivity.* The rate of flow of water through a unit cross-sectional area of an aquifer, at a unit hydraulic gradient. In English units, the rate of flow is in gallons per day per square foot of cross-sectional area. Specific storage or storativity: a measure of the amount of water a confined aquifer will give up for a certain change in head.

2. *Transmissivity.* The rate at which water is transmitted through the whole thickness and unit width of an aquifer under a unit hydraulic gradient. It is equal to the hydraulic conductivity times the thickness of an aquifer.

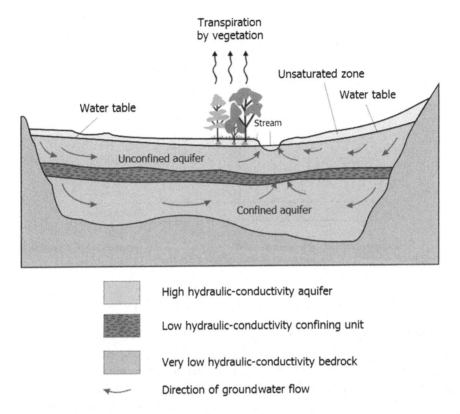

Figure 5.1 Typical flow directions in a cross-sectional view of a simple confined or unconfined aquifer system. The system shows two aquifers with one aquitard a confining or impermeable layer between them, surrounded by the bedrock aquiclude, which is in contact with a gaining stream typical in humid regions (see [110]).

3. *Specific yield or drainable porosity.* A measure of the amount of water an unconfined aquifer will give up when completely drained.

The commonly used groundwater for confined aquifer was proposed by Theis in 1935 [112] from heat transfer literature with the mathematical help of C.I. Lubin, for two-dimensional radial flow to a point source in an infinite, homogeneous aquifer. He noted that when a well-penetrating extensive confined aquifer is pumped at a constant rate, the influence of the discharge extends outward with time. The rate of decline of head, multiplied by the storativity and summed over the area of influence, equals the discharge. The unsteady-state (or Theis) equation, which was derived from the analogy between the flow of groundwater and the conduction of heat, is perhaps the most widely used partial differential equation in groundwater investigations. Figure 5.2 shows the model of the removal of water via a pumping well and the mathematical formula derived by Theis.

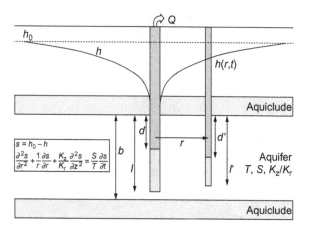

Figure 5.2 The mathematical model of transient flow of water to a pumping well by recognizing the physical analogy between heat flow in solids and groundwater flow in porous media (see [111]).

$$SD_tS(r, t) = TD_{rr}S(r, t) + \frac{1}{r}D_rS(r, t).$$ (5.1)

In the above equation, we have considered the z-component insignificant due to the fact that the movement of water in vertical direction is not important for our investigation. The aforementioned equation is classified under parabolic equation. To include explicitly the variability of the medium through which the flow takes place, the standard version of the partial derivative respect to time is replaced here with beta-derivative to obtain:

$$S_0^A D_t^\beta S(r, t) = TD_{rr}S(r, t) + \frac{1}{r}D_rS(r, t).$$ (5.2)

The following assumptions apply to the use of the Theis-type curve solution:

1. The aquifer has infinite areal extent.
2. The aquifer is homogeneous, isotropic and of uniform thickness.
3. For the one dimensional problem, the pumping well is fully penetrating a confined aquifer
4. The aquifer is nonleaky confined.
5. Flow is unsteady.
6. Water is released instantaneously from storage with the decline of the hydraulic head.
7. Diameter of control well is very small so that storage in the well can be neglected.

5.2.1 Derivation of analytical solution

We devote this section to the derivation of the exact solution of groundwater flow equation with the beta-derivative. We will present the analytical solution using three different techniques, including the beta-Sumudu transform method, the method of separation of variable, and the Botzman method [113]. We shall start with the method of separation of variable. We recall that separation of variables, also known as the Fourier method, is any of several methods for solving ordinary and partial differential equations, in which algebra allows one to rewrite an equation so that each of two variables occurs on a different side of the equation. For a partial differential equation with two parameters, the method assumes that the solution is in the form of [113]:

$$S(r, t) = S_1(r)S_2(t). \qquad (5.3)$$

The above is then replaced in the main equation and two different equations are obtained with inclusion of an Eigen-value. We shall use this method to derive the solution of the new groundwater flow equation. Now, replacing Equation (5.3) into Equation (5.2), we obtain

$$S_1(r)\frac{\partial S_1(t)}{r\partial r} + S_1(r)\frac{\partial^2 S_2(t)}{\partial r^2} = S_2(r){}_0^A D_t^\beta (S_1(t)). \qquad (5.4)$$

Rearranging the above equation, we obtain two separated equation linked with a parameters called Eigen-values and they are provided as:

$$\frac{\partial S_1(r)}{r\partial r} + \frac{\partial^2 S_1(r)}{\partial r^2} = -\lambda^2 S_1(r)\frac{S}{T}{}_0^A D_t^\beta (S_2(t)) = \lambda^2 S_2(t). \qquad (5.5)$$

The spatial ordinary different equation can be solved using the well-known Laplace–Carson transform, defined as:

$$L_c[f(x)](s) = \int_0^\infty x \exp[-xs] \, dx. \qquad (5.6)$$

Properties of this operator can be found in [114]. Now applying the Laplace–Carson transform on both sides of spatial differential equation, we obtain the following

$$-D_s \left[s^2 S_2(s) - sS_2(s) - \frac{dS_2(0)}{dr} \right] + sS_2(s) - S_2(0) - \lambda^2(S_2(s)) = 0. \quad (5.7)$$

Employing the initial conditions, the above equation is reduced to

$$D_s S_2(s)[s^2 + \lambda^2] = -sS_2(s), \qquad (5.8)$$

for which the exact solution is given as

$$S_2(s) = \frac{1}{\sqrt{s^2 + \lambda^2}}. \tag{5.9}$$

Now, applying the inverse Laplace transform operator on both sides in the previous equation, we obtain the following in terms of the Bessel function first kind

$$S_2(r) = J_0(r\lambda) \tag{5.10}$$

$$J_0(r) = \sum_{k=0}^{\infty} \frac{(-1)^k}{k!} \frac{1}{\Gamma(k+1)} \left(\frac{r}{2}\right)^{2k}. \tag{5.11}$$

However, for the second equation, the exact solution is given as:

$$S_2(t) = c \exp\left[-\frac{\lambda T}{\beta S}\left(\left(t + \frac{1}{\Gamma(\beta)}\right)^{\beta} - \left(\frac{1}{\Gamma(\beta)}\right)^{\beta}\right)\right]. \tag{5.12}$$

Thus, using the procedure of the separation of variables, we obtain the exact solution of the new groundwater flow equation as:

$$c \sum_{n=0}^{\infty} \exp\left[-\frac{\lambda_n T}{\beta S}\left(\left(t + \frac{1}{\Gamma(\beta)}\right)^{\beta} - \left(\frac{1}{\Gamma(\beta)}\right)^{\beta}\right)\right] J_0(r\lambda_n). \tag{5.13}$$

Using the initial condition, we obtain the exact solution of the new groundwater equation to be

$$\frac{Q}{4\pi T} \sum_{n=0}^{\infty} \exp\left[-\frac{\lambda_n T}{\beta S}\left(\left(t + \frac{1}{\Gamma(\beta)}\right)^{\beta} - \left(\frac{1}{\Gamma(\beta)}\right)^{\beta}\right)\right] J_0(r\lambda_n). \tag{5.14}$$

We present an alternative method to derive the solution of our equation; this method uses the beta-Laplace transform properties also called Atangana transform [114, 115]. Therefore, using the beta-Laplace transform on both sides of Equation (5.2), we obtain

$$\partial\partial S(r, s) r\partial r + \frac{\partial^2 S(r, s)}{\partial r^2} = \frac{S}{T}(sS(r, s) - S(r, 0)). \tag{5.15}$$

Again, applying the Laplace transform in respect to r, we obtain

$$u\frac{dS(u, s)}{du} + u^2 S(u, s) - u^2 S(0, s) - uS(0, s) - \lambda S(u, s) = 0, \quad \lambda^2 = s\frac{S}{T}. \tag{5.16}$$

Applying the boundary condition together with the initial condition, we obtain the following

$$u\frac{dS(u, s)}{du} + (u^2 - \lambda^2)S(u, s) = 0. \tag{5.17}$$

Applying the boundary condition together with the initial condition, we obtain the following

$$S(u, s) = J_0 \left(su \frac{S}{T} \right). \tag{5.18}$$

Taking double inverse Laplace transform on both sides of above equation yields

$$S(r, t) = c \int_u^\infty \frac{\exp\left[-\frac{T}{S\beta} \lambda^2 \left(\left(t + \frac{1}{\Gamma(\beta)} \right)^\beta - \left(\frac{1}{\Gamma(\beta)} \right)^\beta \right) \right]}{t} \, dt. \tag{5.19}$$

Applying again the initial condition, we obtain the following exact solution of the new groundwater flowing within a confined aquifer

$$S(r, t) = \frac{Q}{4\pi T} \int_u^\infty \frac{\exp\left[-\frac{T}{S\beta} \lambda^2 \left(\left(t + \frac{1}{\Gamma(\beta)} \right)^\beta - \left(\frac{1}{\Gamma(\beta)} \right)^\beta \right) \right]}{t} \, dt$$

$$= \frac{Q}{4\pi T} W_\beta \quad u = \frac{r^2 S}{4Tt}. \tag{5.20}$$

The above derivation can be found in [114]. An alternative derivation of the beta-groundwater flow equation can also be found in [114]. An alternative method is used here to derive the exact solution of the new groundwater flow equation. This method is often used to solve some class of parabolic partial differential equations. This method used the concept of reduction of dimension; in particular, the method used the Boltzmann transformation. In this method, defined for an arbitrary $t_0 < T$ by Equation [114]

$$u_{\beta 0} = \frac{Sr^2}{4T\left[\left(t - t_0 + \frac{1}{\Gamma(\beta)} \right)^\beta - \left(\frac{1}{\Gamma(\beta)} \right)^\beta \right]}. \tag{5.21}$$

Let us consider now the following function [114].

$$S(r, t) = \frac{c}{t - t_0} \exp[-u_{\beta 0}], \tag{5.22}$$

with c being any arbitrary constant. If we assume that r_b is the ratio of the borehole from which the groundwater is being taken out from the aquifer, the total volume of the water withdrawn from the aquifer is provided by:

$$Q_0 \Delta t_0 = 4\pi cT. \tag{5.23}$$

Here

$$S(r, t) = \frac{Q_0 \Delta t_0}{4\pi T(t - t_0)} \exp[-u_{\beta 0}] \qquad (5.24)$$

is the drawdown which will be experimental at a detachment, r, from the pumping well after the time space of Δt_0. Now assume that the above formula is continual m-times, meaning that water is being removed for a very small period of time, Δt_k, at consecutive times $t_{k+1} = t_k + \Delta t_k$, $(k = 0, 1, 2, \ldots, m)$. In this instance, since the new groundwater flow equation is linear, it follows that the total drawdown at any time $t > t_k$ will be given by

$$S(r, t) = \sum_k^n \frac{Q_0 \Delta t_0}{4\pi T(t - t_0)} \exp[-u_{\beta 0}]. \qquad (5.25)$$

Note that, in the above equation, the summation can be transformed into an integral if $\Delta t \to 0$. Then Equation (5.25) becomes

$$S(r, t) = \int_{t_0}^{t} \frac{Q_0 \Delta t_0}{4\pi T(t - x)} \exp[-u_{\beta 0}] \, dx. \qquad (5.26)$$

A particularly important solution arises when t_0 is considered at the origin zero and at the point the discharge rate is independent of time. Equation (5.26) then becomes

$$S(r, t) = \frac{Q}{4\pi T} \int_u^\infty \frac{\exp\left[-\frac{T}{S\beta}\lambda^2\left(\left(t + \frac{1}{\Gamma(\beta)}\right)^\beta - \left(\frac{1}{\Gamma(\beta)}\right)^\beta\right)\right]}{t} \, dt$$

$$= \frac{Q}{4\pi T} W_\beta \quad u = \frac{r^2 S}{4Tt}. \qquad (5.27)$$

the numerical representations of the exact solution of function of time and space for different values of beta.

5.3 STEADY-STATE SOLUTIONS OF THE FLOW IN A CONFINED AND UNCONFINED AQUIFER

In system theory, a system is in a steady state has numerous properties that are unchanging in time. This implies that, for those properties σ of the system, the derivative or partial derivative with respect to time is zero

$$\frac{d\sigma}{dt} = 0 \quad \frac{\partial\sigma}{\partial t} = 0. \qquad (5.28)$$

This concept has relevance in many fields, in particular groundwater flow problems. In this section, we shall present some steady-state solutions of the groundwater flow equation within a confined and unconfined aquifer.

5.3.1 Steady-state solution of the flow in a confined aquifer

Hydraulic conductivity and transmissivity can be determined from steady-state pumping tests; this analysis is called Theim analysis. The following points need to be satisfied for the Theim method to be used.

1. The aquifer is confined.
2. The aquifer has infinite aerial extent.
3. The aquifer is homogeneous, isotropic and of uniform thickness.
4. The piezometric surface is horizontal prior to pumping.
5. The aquifer is pumped at a constant discharge rate.
6. The well penetrates the full thickness of the aquifer and thus receives water by horizontal flow.

The mathematical equation is obtained using Darcy's law and the continuity principle as follows. From Darcy's law we have the following formula

$$q = -K_0^A D_r^\beta (h). \tag{5.29}$$

We next use the continuity concept that says:

$$Q = -2\pi r b q. \tag{5.30}$$

Now, we replace Equation (5.29) into Equation (5.30) to obtain

$$Q = 2\pi r K b_0^A D_r^\beta (h). \tag{5.31}$$

However, applying the beta integral on both side of Equation (5.31) with respect to r:

$$h_2 - h_1 = \frac{Q}{2\pi K b} \int_{r_1}^{r_2} \left(\tau + \frac{1}{\Gamma(\beta)} \right)^{\beta - 1} \frac{1}{\tau} \, d\tau. \tag{5.32}$$

5.3.2 Steady-state solution of the flow in an unconfined aquifer

In the case of an unconfined aquifer, the following assumptions need to be satisfied.

1. The aquifer is unconfined.
2. The aquifer has infinite aerial extent.
3. The aquifer is homogeneous, isotropic and of uniform thickness.
4. The water table is horizontal prior to pumping.
5. The aquifer is pumped at a constant discharge rate.
6. The well penetrates the full thickness of the aquifer and thus receives water from the entire saturated thickness of the aquifer.

The mathematical equation is obtained using Darcy's law and the continuity principle as follows. From Darcy's law we have the following formula

$$q = -K_0^A D_r^\beta (h).$$
(5.33)

We next use the continuity concept, which is different from the one in a confined aquifer:

$$Q = -2\pi r b q.$$
(5.34)

Now, replacing Equation (5.33) into Equation (5.34) to obtain

$$Q = 2\pi r K h b_0^A D_r^\beta (h).$$
(5.35)

However, applying the beta integral on both side of Equation (5.35) with respect to r:

$$\frac{h_2^2 - h_1^2}{2} = \int_{r_1}^{r_2} \left(\tau + \frac{1}{\Gamma(\beta)} \right)^{\beta-1} \frac{1}{\tau} \, d\tau.$$
(5.36)

5.4 MODEL OF GROUNDWATER FLOW EQUATION WITHIN A LEAKY AQUIFER

In nature, leaky aquifers occur far more frequently than the perfectly confined aquifers discussed in the previous chapter. Confining layers overlying or underlying an aquifer are seldom completely impermeable; most of them leak to some extent. When a well in a leaky aquifer is pumped, water is withdrawn not only from the aquifer but also from the overlying and underlying layers. In deep sedimentary basins, it is common for a leaky aquifer to be just one part of a multilayered aquifer system.

The following assumptions apply to the use of the leaky aquifer model for a pumping test solution:

1. The aquifer has infinite areal extent.
2. The aquifer is homogeneous, isotropic and of uniform thickness.
3. The control well is fully or partially penetrating.
4. The flow to control well is horizontal when the control well is fully penetrating.
5. The aquifer is leaky confined.
6. The flow is unsteady.
7. Water is released instantaneously from storage with the decline of the hydraulic head.

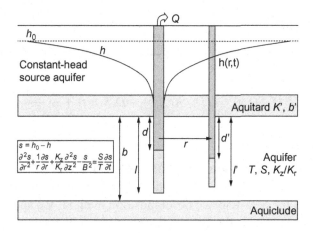

Figure 5.3 The schematic explaining the mathematical model for a leaky confined aquifer pumping test.

8. The diameter of control well is very small, so storage in the well can be neglected.
9. Aquitards have infinite areal extent, uniform vertical hydraulic conductivity and uniform thickness.
10. Aquitards are overlain or underlain by an infinite constant-head plane source.
11. Aquitards are incompressible (no storage).
12. The flow in the aquitards is vertical.

Figure 5.3 shows and explains the model of the well-aquifer configuration for a pumping test in a leaky confined aquifer. When a leaky aquifer is pumped, the piezometric level of the aquifer in the well is lowered. This lowering spreads radically outward as pumping continues, creating a difference in hydraulic head between the aquifer and the aquitards. Consequently, the groundwater in the aquitards will start moving vertically downward to join the water in the aquifer. The aquifer is thus partially recharged by downward percolation from the aquitards. As pumping continues, the percentage of the total discharge derived from this percolation increases. After a certain period of pumping, equilibrium will be established between the discharge rate of the pump and the recharge rate by vertical flow through the aquitards. This steady state will be maintained as long as the water table in the aquitards is kept constant. Hantush was the first to derive a partial differential equation describing such phenomena. According to Hantush and Jacob [115, 116], the drawdown due to pumping a leaky aquifer can be described by the following equation:

$$SD_tS(r,t) = TD_{rr}S(r,t) + \frac{1}{r}D_rS(r,t) + \frac{S(r,t)}{\lambda^2}. \tag{5.37}$$

In Equation (5.37), we replace the local derivative with the beta-derivative to obtain

$$S_0^A D_t^\beta S(r,t) = TD_{rr}S(r,t) + \frac{1}{r}D_rS(r,t) + \frac{S(r,t)}{\lambda^2}. \tag{5.38}$$

5.4.1 Special solution via iteration method

To derive the special solution to Equation (5.38), we make use of an iterative method called the beta-Laplace decomposition method. We applied on both sides of Equation (5.38) the beta-Laplace transform to obtain:

$$L_\beta S_0^A D_t^\beta S(r,t) = L_\beta\left(TD_{rr}S(r,t) + \frac{1}{r}D_rS(r,t) + \frac{S(r,t)}{\lambda^2}\right)$$

$$s(S(r,s) - S(r,0)) = L_\beta\left(TD_{rr}S(r,t) + \frac{1}{r}D_rS(r,t) + \frac{S(r,t)}{\lambda^2}\right)$$

$$S(r,s) - S(r,0) = \frac{1}{s}L_\beta\left(TD_{rr}S(r,t) + \frac{1}{r}D_rS(r,t) + \frac{S(r,t)}{\lambda^2}\right). \tag{5.39}$$

Applying the inverse Laplace transform operator on both sides of Equation (5.39), we obtain:

$$S(r,t) - S(r,0) = L^{-1}\left(\frac{1}{s}L_\beta\left(TD_{rr}S(r,t) + \frac{1}{r}D_rS(r,t) + \frac{S(r,t)}{\lambda^2}\right)\right). \tag{5.40}$$

Now, we assume that the solution of Equation (5.40) is in the form of

$$S(r,t) = \sum_{n=0}^{\infty} p^n S_n(r,t). \tag{5.41}$$

Replacing Equation (5.41) into Equation (5.42), we obtain the following

$$\sum_{n=0}^{\infty} p^n S_n(r,t) - S(r,0) = pL^{-1}\left(\frac{1}{s}L_\beta\left(TD_{rr}\left(\sum_{n=0}^{\infty} p^n S_n(r,t)\right)\right.\right.$$

$$\left.\left. + \frac{1}{r}D_r\left(\sum_{n=0}^{\infty} p^n S_n(r,t)\right) + \frac{\left(\sum_{n=0}^{\infty} p^n S_n(r,t)\right)}{\lambda^2}\right)\right). \tag{5.42}$$

We now compare terms of same power of p to obtain

$$S_0(r,t) = S(r,0)$$

$$S_1(r,t) = L^{-1}\left(\frac{1}{s}L_\beta\left(TD_{rr}S_0(r,t) + \frac{1}{r}D_rS_0(r,t) + \frac{S_0(r,t)}{\lambda^2}\right)\right)$$

$$S_2(r,t) = L^{-1}\left(\frac{1}{s}L_\beta\left(TD_{rr}S_1(r,t) + \frac{1}{r}D_rS_1(r,t) + \frac{S_1(r,t)}{\lambda^2}\right)\right)$$

$$S_3(r,t) = L^{-1}\left(\frac{1}{s}L_\beta\left(TD_{rr}S_2(r,t) + \frac{1}{r}D_rS_2(r,t) + \frac{S_2(r,t)}{\lambda^2}\right)\right)$$

$$\vdots$$

$$S_n(r,t) = L^{-1}\left(\frac{1}{s}L_\beta\left(TD_{rr}S_{n-1}(r,t) + \frac{1}{r}D_rS_{n-1}(r,t) + \frac{S_{n-1}(r,t)}{\lambda^2}\right)\right).$$

$$(5.43)$$

5.4.2 Stability and convergence analysis

The stability analysis of the used method is achieved by employing some properties of inner product within a constructed Banach space $\mathcal{H}$ defined as $\mathcal{H} = \left(S, V/\int SV < \infty\right)$. We next consider the following operator

$$H(S) = SD_tS(r,t) = TD_{rr}S(r,t) + \frac{1}{r}D_rS(r,t) + \frac{S(r,t)}{\lambda^2}. \qquad (5.44)$$

We aim here to evaluate the following expression where (.) is the inner product:

$$(H(S) - H(V), S - V). \qquad (5.45)$$

By definition, we have

$$H(S) - H(V) = TD_{rr}(S(r,t) - V(r,t)) + \frac{1}{r}D_r(S(r,t) - V(r,t)) + \frac{(S(r,t) - V(r,t))}{\lambda^2}. \qquad (5.46)$$

Therefore, using the linear property of the inner product, we obtain the following equation

$$(H(S) - H(V), S - V) = (TD_{rr}(S(r,t) - V(r,t)), S(r,t) - V(r,t))$$
$$+ \left(\frac{1}{r}D_r(S(r,t) - V(r,t)), S(r,t) - V(r,t)\right)$$
$$+ \left(\frac{(S(r,t) - V(r,t))}{\lambda^2}, S(r,t) - V(r,t)\right). \qquad (5.47)$$

Using the link between the inner product and the norm, case by case, we evaluate first,

$$
\begin{aligned}
(TD_{rr}(S(r,t)-V(r,t)), S(r,t)-V(r,t)) &\le T\|D_{rr}(S(r,t)-V(r,t))\|\,\|S(r,t)-V(r,t)\| \\
&\le T\omega_1 \|D_r(S(r,t)-V(r,t))\|\,\|S(r,t)-V(r,t)\| \\
&\le T\omega_1\omega_2 \|S(r,t)-V(r,t)\|^2. \quad (5.48)
\end{aligned}
$$

Also,

$$
\begin{aligned}
\left(\frac{1}{r}D_r(S(r,t)-V(r,t)), S(r,t)-V(r,t)\right) &\le \max_{r\in \text{aquifer}} [r^{-1}]\|D_r(S(r,t)-V(r,t))\| \\
&\quad \|S(r,t)-V(r,t)\| \\
&\le \max_{r\in \text{aquifer}} [r^{-1}]\omega_3 \|S(r,t)-V(r,t)\|^2.
\end{aligned}
$$

$$(5.49)$$

Lastly,

$$
\left(\frac{(S(r,t)-V(r,t))}{\lambda^2}, S(r,t)-V(r,t)\right) \le \frac{1}{\lambda^2}\|S(r,t)-V(r,t)\|^2. \quad (5.50)
$$

Replacing Equations (5.50), (5.49), and (5.48) into Equation (5.47), we obtain

$$
\begin{aligned}
(H(S)-H(V), S-V) &\le \left(T\omega_1\omega_2 + \omega_3 \max_{r\in \text{aquifer}} (r^{-1}) + \frac{1}{\lambda^2}\right)\|S(r,t)-V(r,t)\|^2 \\
&\le K\|S(r,t)-V(r,t)\|^2. \quad (5.51)
\end{aligned}
$$

We next evaluate the following expression

$$
\begin{aligned}
(H(S)-H(V), W(r,t)) &= (TD_{rr}(S(r,t)-V(r,t)), W(r,t)) \\
&\quad + \left(\frac{1}{r}D_r(S(r,t)-V(r,t)), W(r,t)\right) \\
&\quad + \left(\frac{(S(r,t)-V(r,t))}{\lambda^2}, W(r,t)\right). \quad (5.52)
\end{aligned}
$$

Using the link between the inner product and the norm, case by case, we evaluate first,

$$
\begin{aligned}
(TD_{rr}(S(r,t)-V(r,t)), W(r,t)) &\le T\|D_{rr}(S(r,t)-V(r,t))\|\,\|W(r,t)\| \\
&\le T\gamma_1 \|D_r(S(r,t)-V(r,t))\|\,\|W(r,t)\| \\
&\le T\gamma_1\gamma_2 \|S(r,t)-V(r,t)\|\,\|W(r,t)\|.
\end{aligned}
$$

$$(5.53)$$

Also,

$$\left(\frac{1}{r}D_r(S(r,t)-V(r,t)), W(r,t)\right) \leq \max_{r\in \text{ aquifer}} [r^{-1}]\|D_r(S(r,t)-V(r,t))\|\,\|W(r,t)\|$$

$$\leq \max_{r\in \text{ aquifer}} [r^{-1}]\gamma_3\|S(r,t)-V(r,t)\|\,\|W(r,t)\|.$$

(5.54)

Lastly,

$$\left(\frac{(S(r,t)-V(r,t))}{\lambda^2}, S(r,t)-V(r,t)\right) \leq \frac{1}{\lambda^2}\|S(r,t)-V(r,t)\|\,\|W(r,t)\|.$$

(5.55)

Replacing Equations (5.55), (5.54), and (5.53) into Equation (5.52), we obtain

$$(H(S)-H(V), W) \leq \left(T\gamma_1\gamma_2+\gamma_3 \max_{r\in \text{ aquifer}} (r^{-1})+\frac{1}{\lambda^2}\right)\|S(r,t)-V(r,t)\|\,\|W(r,t)\|$$

$$\leq M\|S(r,t)-V(r,t)\|\,\|W(r,t)\|.$$

(5.56)

Indeed, Equations (5.56) and (5.51) complete the proof of stability of the used method.

5.4.3 Uniqueness analysis of the special solution

Here, we present the unicity of the special solution while using the iterative method. To achieve this, we assume that S is the exact solution of our equation. Let S_1 and S_2 two different special solutions, such that for we can find m_1 and m_2 for after which both converge to the exact solution S, then, for all $W \in \mathcal{H}$

$$(H(S_1)-H(S_2), W(r,t)) = (TD_{rr}(S_1(r,t)-S_2(r,t)), W(r,t))$$
$$+\left(\frac{1}{r}D_r(S_1(r,t)-S_2(r,t)), W(r,t)\right)$$
$$+\left(\frac{(S_1(r,t)-S_2(r,t))}{\lambda^2}, W(r,t)\right).$$

(5.57)

Using the link between the inner product and the norm, case by case, we evaluate first,

$$(TD_{rr}(S_1(r,t)-S_2(r,t)), W(r,t)) \leq T\|D_{rr}(S_1(r,t)-S_2(r,t))\|\,\|W(r,t)\|$$
$$\leq T\zeta_1\|D_r(S_1(r,t)-S_2(r,t))\|\,\|W(r,t)\|$$

$$\leq T\zeta_1\zeta_2\|S(r,t)-V(r,t)\|\,\|W(r,t)\|.$$

$$(5.58)$$

Also,

$$\left(\frac{1}{r}D_r(S_1(r,t)-S_2(r,t)),W(r,t)\right) \leq \max_{r\in\text{ aquifer}}[r^{-1}]\|D_r(S_1(r,t)-S_2(r,t))\|\,\|W(r,t)\|$$

$$\leq \max_{r\in\text{ aquifer}}[r^{-1}]\zeta_3\|S_1(r,t)-S_2(r,t)\|\,\|W(r,t)\|.$$

$$(5.59)$$

Lastly,

$$\left(\frac{(S_1(r,t)-S_2(r,t))}{\lambda^2},S_1(r,t)-S_2(r,t)\right) \leq \frac{1}{\lambda^2}\|S_1(r,t)-S_2(r,t)\|\,\|W(r,t)\|.$$

$$(5.60)$$

Replacing Equations (5.60), (5.59), and (5.58) into Equation (5.57), we obtain

$$(H(S)-H(V),W) \leq \left(T\zeta_1\zeta_2+\zeta_3\max_{r\in\text{ aquifer}}(r^{-1})+\frac{1}{\lambda^2}\right)\|S(r,t)-S_2(r,t)\|\,\|W(r,t)\|$$

$$\leq B\|S_1(r,t)-S_2(r,t)\|\,\|W(r,t)\|.$$

$$(5.61)$$

However, the right-hand side of Equation (5.61) can be reformulated as follows

$$B\|S_1(r,t)-S_2(r,t)\|\,\|W(r,t)\| = B\|S_1(r,t)+S(r,t)-S(r,t)-S_2(r,t)\|\,\|W(r,t)\|$$

$$\leq \|S_1(r,t)-S(r,t)\|+\|S_2(r,t)-S(r,t)\|. \quad (5.62)$$

Nevertheless, using the fact that, we can find m_1 and m_2 for after which both converge to the exact solution S,

$$\|S_1(r,t)-S(r,t)\| < \frac{\epsilon}{2B\|W(r,t)\|}, \quad \|S_2(r,t)-S(r,t)\| < \frac{\epsilon}{2B\|W(r,t)\|}.$$

$$(5.63)$$

Now taking $m=\max(m_1,m_2)$, we obtain

$$B\|S_1(r,t)-S_2(r,t)\| < \epsilon. \quad (5.64)$$

Employing idea of boundness in topology, we conclude that

$$\|S_1(r,t)-S_2(r,t)\| = 0 \Longleftrightarrow S_1(r,t)=S_2(r,t)\forall(r,t)\in \text{Dom}. \quad (5.65)$$

This conclude the proof of uniqueness.

5.4.4 Numerical simulation

An aquifer test (or a pumping test) is conducted to evaluate an aquifer by "stimulating" the aquifer through constant pumping, and observing the aquifer's *response or drawdown* in observation wells. A slug test is a variation on the typical aquifer test where an instantaneous change (increase or decrease) is made, and the effects are observed in the same well. Aquifer tests are typically interpreted by using an analytical model of aquifer flow to match the data observed in the real world, then assuming that the parameters from the idealized model apply to the real-world aquifer. In more complex cases, a numerical model may be used to analyze the results of an aquifer test, but adding complexity does not ensure better results. Typically monitoring and pumping wells are screened across the same aquifers. Most commonly, an aquifer test is conducted by pumping water from one well at a steady rate and for at least 1 day, while carefully measuring the water levels in the monitoring wells. When water is pumped from the pumping well, the pressure in the aquifer that feeds that well declines. This decline in pressure will show up as drawdown (change in the hydraulic head) in an observation well. Drawdown decreases with radial distance from the pumping well, and drawdown increases with the length of time that the pumping continues. The aquifer characteristics that are evaluated by most aquifer tests are:

1. *Hydraulic conductivity.* The rate of flow of water through a unit cross-sectional area of an aquifer, at a unit hydraulic gradient. In English units,

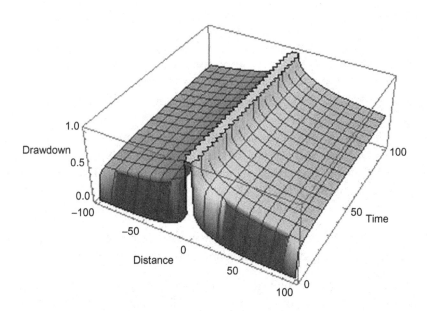

Figure 5.4 Numerical simulation of the beta-groundwater flow equation solution as a function of space and time for $\beta = 0.5$.

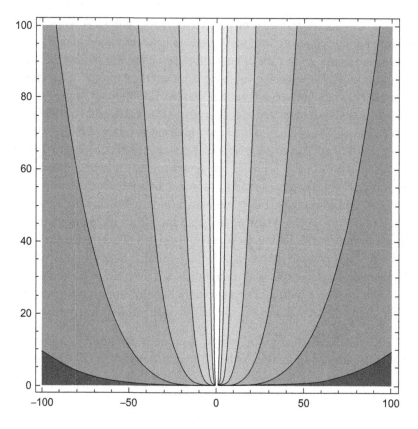

Figure 5.5 Contour plot of the numerical simulation for solution of the beta-groundwater flow equation as function of time and space for $\beta = 0.5$.

the rate of flow is in gallons per day per square foot of cross-sectional area.

2. *Specific storage or storativity.* A measure of the amount of water that a confined aquifer will give up for a certain change in head.

3. *Transmissivity.* The rate at which water is transmitted through whole thickness and unit width of an aquifer under a unit hydraulic gradient. It is equal to the hydraulic conductivity times the thickness of an aquifer.

We present in Figures 5.4 and 5.5 the numerical replication of the beta-groundwater flow equation as a function of time and space for $\beta = 0.5$.

Note that a contour plot gives you essentially a typography map of a function. The contours join points on the surface that have the same height. The default is to have contours corresponding to a sequence of equally spaced z-values.

5.5 MODEL OF LASSA FEVER OR LASSA HEMORRHAGIC FEVER

Lassa fever or Lassa hemorrhagic fever (LHF) is an acute viral hemorrhagic fever caused by the Lassa virus. It was first described in 1969 in the town of Lassa, in Borno State, Nigeria [117]. Lassa fever is a member of the Arenaviridae virus family. Similar to Ebola [107], clinical cases of the disease had been known for over a decade but had not been connected with a viral pathogen. The infection is endemic in West African countries, resulting in 300,000–500,000 cases annually, causing approximately 5000 deaths each year [118]. Outbreaks of the disease have been observed in Nigeria, Liberia, Sierra Leone, Guinea, and the Central African Republic. The primary animal host of the Lassa virus is the natal multimammate mouse (*Mastomys natalensis*), an animal indigenous to most of Sub-Saharan Africa [119]. Figure 5.6 shows the host of the Lassa fever virus. The virus is probably transmitted by contact with the feces or urine of animals accessing grain stores in residences [120]. Given its high rate of incidence, Lassa fever is a major problem in affected countries. Lassa virus is zoonotic (transmitted from animals), in that it spreads to humans from rodents, specifically multimammate rats (*M. natalensis*) [120]. This is probably the most common rodent in equatorial Africa, ubiquitous in human households and eaten as a delicacy in some areas [121]. In these rats, infection is in a persistent asymptomatic state. The virus is shed in their excreta (urine and feces), which can be aerosolized. In fatal cases, Lassa fever is characterized by impaired or delayed cellular immunity leading to fulminant viremia (Figure 5.7).

Figure 5.6 The natal multimammate mouse (M. natalensis).

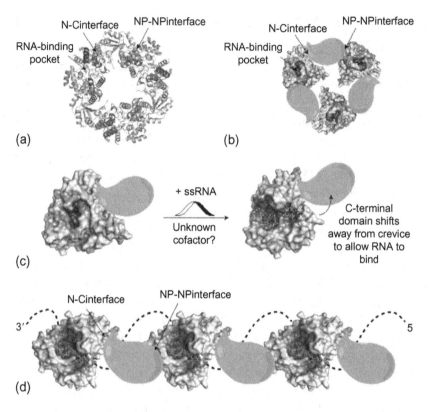

Figure 5.7 A model for arenavirus RNP organization. (A) Organization of the trimeric, RNA-free LASV NP. (B) The N-terminal domain of LASV NP colored by electrostatic surface potential and the C-terminal domain modeled as a cartoon (green) demonstrate the "closed" form of the NP structure. In this conformation, the RNA-binding crevice is not available to accept ssRNA. (C) To bind the viral genome, the C-terminal domain must shift away from the RNA-binding crevice to allow RNA to enter. This could be initiated by binding of NP by an as-yet-identified cofactor or perhaps the viral genome itself. (D) When bound to ssRNA, the trimer of NP will not form. Instead monomers of NP line the ssRNA backbone. Each N-terminal domain of NP interacts with the adjacent C-terminal domain of a neighboring NP.

Infection in humans typically occurs by exposure to animal excrement through the respiratory or gastrointestinal tracts. Inhalation of tiny particles of infectious material (aerosol) is believed to be the most significant means of exposure. It is possible to acquire the infection through broken skin or mucous membranes that are directly exposed to infectious material. Transmission from person to person has also been established, presenting a disease risk for healthcare workers. Frequency of transmission via sexual contact has not been established. In 80% of cases, the disease is asymptomatic, but in the remaining 20%, it takes a complicated course. As noted above, it is estimated that the virus is responsible for about 5000 deaths annually. The fever accounts for up to one-third of deaths in hospitals within the affected regions and 10–16% of total cases [107, 118, 119, 123]. After an incubation period of 6–21 days, an acute illness with multiorgan involvement develops. Nonspecific symptoms include fever, facial swelling, and muscle fatigue, as well as conjunctivitis and mucosal bleeding. The other symptoms arising from the affected organs are:

1. gastrointestinal tract nausea;
2. vomiting (bloody);
3. diarrhea (bloody);
4. stomach ache;
5. constipation;
6. dysphagia (difficulty swallowing);
7. hepatitis;
8. cardiovascular system pericarditis;
9. hypertension;
10. hypotension;
11. tachycardia (abnormally high heart rate);
12. respiratory tract cough;
13. chest pain;
14. dyspnoea;
15. pharyngitis;
16. pleuritis;
17. nervous system encephalitis;
18. meningitis;
19. unilateral or bilateral hearing deficit; and
20. seizures.

Clinically, Lassa fever infections are difficult to distinguish from other viral hemorrhagic fevers such as Ebola and Marburg, and from more common febrile illnesses such as malaria. The virus is excreted in urine for 3–9 weeks and in semen for 3 months. A mathematical model using the beta-derivative describing the spread of this disease in pregnant women was proposed by Abdon Atangana in 2015 [124]. We present in the next section the results obtained from his work.

5.5.1 Mathematical model of Lassa using the beta-derivative

Let N be a total number of adult women in a given country, S be the susceptible population of pregnant women, R be the recovery population of pregnant women, I the infected population of pregnant women, and D the population of pregnant women dying in that country. We shall assume that women are being pregnant at the rate b, they are susceptible at a rate a, they are infected at a rate c, the infected women are dying at a rate f, or recovering at the rate h. We assume that they die of natural causes or other disease at a rate l. Then the mathematical formula underpinning the change in time of the susceptible population within the scope of beta-derivative is given, as [124]:

$$ {}_0^A D_t^\beta S(t) = -cS(t)I(t) + bN(t) - lN + fR(t) - lS(t). \tag{5.66} $$

Equation (5.66) is obtained because c is the rate of infectious pregnant women from recovery population; they turn out to be vulnerable again at the rate h; a proportion of adult women will be pregnant at a rate b; and finally a number of pregnant women die due to natural causes and other diseases at the rate l. The change of infected populations can be expressed with the following linear ordinary differential equation:

$$ {}_0^A D_t^\beta = cS(t)I(t) - (f+h)I(t) - hS(t). \tag{5.67} $$

The physical explanation underpinning the above equation is that the total number of pregnant women removed from susceptible group can be mathematically expressed as $cS(t)I(t)$ [114]. However, due to the introduction of medication, a number of pregnant women will recover at a rate of h, and a number of pregnant women will die at a rate f. The change in time of the recovery population is given by

$$ {}_0^A D_t^\beta R(t) = hI(t) - fR(t). \tag{5.68} $$

Finally, we can express the change in time of the population of the death as

$$ {}_0^A D_t^\beta D(t) = fI(t) + lN - bN + lS(t). \tag{5.69} $$

Therefore, the description of the spread and consequences associate can be underpinned by the following set of mathematical formula [114]:

$$ \begin{cases} {}_0^A D_t^\beta S(t) = -cS(t)I(t) + bN(t) - lN + fR(t) - lS(t) \\ {}_0^A D_t^\beta I(t) = cS(t)I(t) - (f+h)I(t) - hS(t) \\ {}_0^A D_t^\beta R(t) = hI(t) - fR(t) \\ {}_0^A D_t^\beta D(t) = fI(t) + lN - bN + lS(t). \end{cases} \tag{5.70} $$

5.5.2 Analysis of equilibrium points

The basic reproductive number R_0 is typically defined [125] as the average number of secondary cases produced by a typical infected individual during his/her entire time of being infectious when introduced in a population of susceptibles. This nondimensional quantity cannot be computed explicitly in most cases because the mathematical description of what is a typical infectious individual is difficult to quantify in populations with a high degree of heterogeneity. Regardless of whether or not $\mathcal{R}_0$ can be computed explicitly, its role on the study of the stability of equilibria can still be determined. Most reasonable epidemic models support at least two types of equilibria: a disease-free equilibria and a positive (endemic) equilibria. Typically, one can show that the disease-free equilibrium is locally asymptotically stable if $R_0 < 1$ and unstable if $\mathcal{R}_0 > 1$. Furthermore, in many examples, it has been shown that $\mathcal{R}_0 > 1$ implies the existence of a unique endemic equilibrium. Many models found in the literature have been used to show that when $\mathcal{R}_0$ crosses the threshold, $\mathcal{R}_0 = 1$, a transcritical bifurcation takes place. That is, asymptotic local stability is transferred from the infectious-free state to the new (emerging) endemic (positive) equilibria. In some situations, it can be shown that the transfer of asymptotic stability is independent of initial conditions; that is, it is global [126]. An alternative method to study the stability of the disease equilibrium point called the "next generation operator approach" was introduced in 1990 [127]. In their work, the authors defined $\mathcal{R}_0$ as the spectral radius of the next generation operator. The details of this approach are outlined in the rest of this section. First, we consider the case where heterogeneity is discrete, being defined using groups defined by fixed characteristics, for epidemiological models that can be written in the form:

$$\begin{cases} \frac{dX}{dt} = f(X, Y, Z), \\ \frac{dY}{dt} = g(X, Y, Z), \\ \frac{dZ}{dt} = h(X, Y, Z), \end{cases} \tag{5.71}$$

where $X \in \mathcal{R}^r$, $Y \in \mathcal{R}^s$, $Z \in \mathcal{R}^n$, $r, s, n \geq 0$, and $h(X, 0, 0) = 0$. The components of X denote the number of susceptibles, recovered, and other classes of noninfected individuals. The components of Y represent the number of infected individuals who do not transmit the disease or various latent or noninfectious stages. The components of Z represent the number of infected individuals capable of transmitting the disease. Let $V_0 = (X^*, 0, 0) \in \mathcal{R}^{s+n+r}$ denote the disease-free equilibrium, which means

$$f(X^*, 0, 0) = g(X^*, 0, 0) = h(X^*, 0, 0) = 0. \tag{5.72}$$

Assuming that the equation

$$g(X^*, Y, Z) = 0 \tag{5.73}$$

implicitly determines a function

$$Y = \bar{g}(X^*, Y). \tag{5.74}$$

Let $A = D_z h(X^*, \bar{g}(X^*, 0), 0)$ and further assume that A can be written in form $A = M - D$, with $M \geq 0$ (i.e., $m_{ij} \geq 0$) and $D > 0$, a diagonal matrix. The spectral bound of matrix B is denoted by $m(B) = \sup R\lambda : \lambda \in \sigma(B)$, where $R\lambda$ denotes the real part of λ, while $\rho(B) = \lim_{n \to \infty} \|B^n\|^{\frac{1}{n}}$ denotes the spectral radius of B, the proof of involving matrix A can be found in [127].

Theorem 5.5.1.

$$m(A) < 0 \iff \rho(MD^{-1}) < 1, \tag{5.75}$$

or

$$m(A) > 0 \iff \rho(MD^{-1}) > 1. \tag{5.76}$$

The basic reproductive number is defined as the spectral radius a dominant Eigen-value of the matrix MD^{-1}, which means

$$\mathcal{R}_0 = \rho(MD^{-1}). \tag{5.77}$$

An analogous formula for $\mathcal{R}_0$ when a heterogenous population is stratified by continuous characteristics [126] can be similarly computed. In fact, let $S(\eta)$ denotes the population density function that describes the steady state in the absence of disease where $\eta \in \Omega_h$ where h represents heterogeneity. In addition, let $A(\tau, \eta, \zeta)$ denote the current expected infectivity of an individual who was infected τ units of time ago while at state ζ; that is, $A(\tau, \eta, \zeta)$ denotes the average infectivity that can be exercised on an uninfected individual at state η, provided the that uninfected population finds itself at the steady demographic state $S(\eta)$. The function $A(\tau, \eta, \zeta)$ combines information on the probability per unit of time that contacts between certain stages take place and the probability that, given a contact, the disease agent is actually transmitted. Under the special assumption of proportionate mixing [128], $A(\tau, \eta, \zeta)$ can be expressed in the form

$$A(\tau, \eta, \zeta) = f(\eta)g(\tau, \zeta). \tag{5.78}$$

$\mathcal{R}_0$, the spectral radius of the next generation operator, can be computed under proportionate mixing. In fact, it is given by the following formula:

$$\mathcal{R}_0 = \int_\Omega \int_0^\infty g(\tau, \zeta)S(\eta) \, d\tau \, d\zeta. \tag{5.79}$$

The key element, in the computation of $\mathcal{R}_0$ in Equation (5.79), is the infectivity function $A(\tau, \eta, \zeta)$.

5.5.3 Application to model of Lassa fever with beta-derivative

To find the endemic equilibrium points, we assume that the system is time independent such as that using one of the properties of the beta-derivative we have [114]:

$$
\begin{cases}
0 = -c\overline{SI} + bN - l\overline{N} + f\overline{R} - \overline{IS} \\
0 = c\overline{SI} - (f+h)\overline{I} - h\overline{S} \\
0 = h\overline{I} - f\overline{R} \\
0 = f\overline{I} + l\overline{N} - bN + \overline{IS}.
\end{cases}
\tag{5.80}
$$

By solving the last two equations of the system, we obtain

$$
\overline{R} = \frac{h\overline{I}}{f}, \quad \overline{S} = \frac{f+h}{c\overline{I} - h}.
\tag{5.81}
$$

Now, replacing the above solutions into the first equation of the system, we obtain

$$
\begin{cases}
A\overline{I}^2 + B\overline{I} + C = 0, \quad C = -h(b - l)N, \\
B = (h + c - l)f + cN(b - l) + h(c - l), \quad A = hc
\end{cases}
.
\tag{5.82}
$$

Thus the solution of the above equation is given as

$$
\overline{I}_+ = \frac{-B + \sqrt{B^2 + Chc}}{2A}, \quad \overline{I}_- = \frac{-B - \sqrt{B^2 + Chc}}{2A}.
\tag{5.83}
$$

We consider only the positive solution and obtain the endemic equilibrium points.

$$
\overline{I} = \frac{\sqrt{B^2 + 4Chc}}{2A} = \frac{B}{2A}\left(\frac{\sqrt{B^2 + 4Chc}}{B} - 1\right)
$$

$$
= \frac{B}{2A}(\mathcal{R}_0 - 1).
\tag{5.84}
$$

It is worth noting that a large value of $\mathcal{R}_0$ may indicate the possibility of a major epidemic. In our case, the reproductive number is given as

$$
\mathcal{R}_0 = \frac{\sqrt{B^2 + 4Chc}}{B}.
\tag{5.85}
$$

Accordingly, if B is negative, then $\mathcal{R}_0 < 1$ and the disease-free equilibrium is stable and the endemic equilibrium points are unstable. If B is positive and *Chc* is positive, then $\mathcal{R}_0 > 1$, the disease-free equilibrium is unstable and the endemic stable [114].

5.5.4 Special analytical solution via an iterative method with Atangana's transform

One of the key aspects of modeling is perhaps the simulation or the prediction of the physical problem using the mathematical formula. In order to achieve this, we solve the proposed mathematical formula numerically or analytically. Whether it is numerical or analytical, when we are dealing with nonlinear equations, the problem become more demanding. There are quite few methods in the literature dealing with nonlinear equations [114]. Now, making use of the above operator, on both sides of Equation (5.70), we obtain

$$
\begin{cases}
S(s) = \frac{S(0)}{s} \frac{1}{s}\mathcal{L}_\beta\left(-cS(t)I(t) + bN(t) - lN + fR(t) - lS(t)\right) \\
I(s) = \frac{I(0)}{s} + \frac{1}{s}\mathcal{L}_\beta\left(cS(t)I(t) - (f+h)I(t) - hS(t)\right) \\
R(s) = \frac{R(0)}{s} + \frac{1}{s}\mathcal{L}_\beta\left(hI(t) - fR(t)\right) \\
D(s) = \frac{D(0)}{s} + \frac{1}{s}\mathcal{L}_\beta\left(fI(t) + lN - bN + lS(t)\right).
\end{cases}
\tag{5.86}
$$

Thus, applying the inverse Laplace operator on both sides of the above, we obtain

$$
\begin{cases}
S(t) = S(0) + \mathcal{L}\left(\frac{1}{s}\mathcal{L}_\beta\left(-cS(t)I(t) + bN(t) - lN + fR(t) - lS(t)\right)\right) \\
I(t) = I(0) + \mathcal{L}\left(\frac{1}{s}\mathcal{L}_\beta\left(cS(t)I(t) - (f+h)I(t) - hS(t)\right)\right) \\
R(t) = R(0) + \mathcal{L}\left(\frac{1}{s}\mathcal{L}_\beta\left(hI(t) - fR(t)\right)\right) \\
D(t) = D(0) + \mathcal{L}\left(\frac{1}{s}\mathcal{L}_\beta\left(fI(t) + lN - bN + lS(t)\right)\right).
\end{cases}
\tag{5.87}
$$

The iterative method can now be employed to put frontward the main recursive formula connecting the Lagrange multiplier as

$$
\begin{cases}
S_0(t) = S(0) \\
I_0(t) = I(0) \\
R_0(t) = R(0) \\
D_0(t) = D(0),
\end{cases}
\tag{5.88}
$$

with

$$
\begin{cases}
S_{n+1}(t) = S_n(t) + \mathcal{L}\left(\frac{1}{s}\mathcal{L}_\beta\left(-cS_n(t)I_n(t) + bN(t) - lN + fR_n(t) - lS_n(t)\right)\right) \\
I_{n+1}(t) = I_n(t) + \mathcal{L}\left(\frac{1}{s}\mathcal{L}_\beta\left(cS_n(t)I_n(t) - (f+h)I_n(t) - hS_n(t)\right)\right) \\
R_{n+1}(t) = R_n(t) + \mathcal{L}\left(\frac{1}{s}\mathcal{L}_\beta\left(hI_n(t) - fR_n(t)\right)\right) \\
D_{n+1}(t) = D_n(t) + \mathcal{L}\left(\frac{1}{s}\mathcal{L}_\beta\left(fI_n(t) + lN - bN + lS_n(t)\right)\right).
\end{cases}
\tag{5.89}
$$

The special solution of this equation is therefore given as

$$\begin{cases} S(t) = \lim_{n\to\infty} S_n(t) \\ I(t) = \lim_{n\to\infty} I_n(t) \\ R(t) = \lim_{n\to\infty} R_n(t) \\ D(t) = \lim_{n\to\infty} D_n(t). \end{cases} \tag{5.90}$$

5.5.5 Stability and unicity analysis for the iteration method

The efficiency of the used method can only be expressed via the stability and the convergence analysis. Therefore, we present in this section the stability analysis of the used method for solving the novel system Equation (5.91). To achieve this, we consider the following operator

$$\begin{cases} {}_0^A D_t^\beta S(t) = -cS(t)I(t) + bN(t) - lN + fR(t) - lS(t) \\ {}_0^A D_t^\beta I(t) = cS(t)I(t) - (f+h)I(t) - hS(t) \\ {}_0^A D_t^\beta R(t) = hI(t) - fR(t) \\ {}_0^A D_t^\beta D(t) = fI(t) + lN - bN + lS(t) \end{cases} \tag{5.91}$$

$$O(S,I,R,D) = \begin{cases} {}_0^A D_t^\beta S(t) = cS(t)I(t) + bN(t) - lN + fR(t) - lS(t) \\ {}_0^A D_t^\beta I(t) = cS(t)I(t) - (f+h)I(t) - hS(t) \\ {}_0^A D_t^\beta R(t) = hI(t) - fR(t) \\ {}_0^A D_t^\beta D(t) = fI(t) + lN - bN + lS(t). \end{cases}$$
$$\tag{5.92}$$

Theorem 5.5.2. *Let us consider the above operator O and think about the initial condition for the system of Equation (5.91), then the method used leads to a special solution of system [114].*

Proof. We consider the following Z-sub-Hilbert space of the Hilbert space $\mathcal{H} = L^2((0,T))$, that can be defined as the set of those functions of

$$P : (0,T) \longrightarrow \mathcal{R}, \quad Z = \left(u, v \int_0^t \left(\tau + \frac{1}{\Gamma(\beta)} \right)^{\beta-1} uv < \infty \right). \tag{5.93}$$

We agreeably undertake that the differential operators are limited under the L^2 norms. Exploiting the description of the operator, O, we ensure the succeeding [114]

$$O(S,I,R,D) - O(S_1,I_1,R_1,D_1)$$

$$\begin{cases} cS(t)I(t) - cS_1(t)I_1(t) + f(R(t) - R_1(t)) - l(S(t) - S_1(t)) \\ cS(t)I(t) - cS_1(t)I_1(t) - (f + h)(I(t) - I_1(t)) - h(S(t) - S_1(t)) \\ h(I(t) - I_1(t)) - f(R(t) - R_1(t)) \\ f(I(t) - I_1(t)) + l(S(t) - S_1(t)). \end{cases}$$

$$(5.94)$$

We shall evaluate next the beta inner product of

$$G = A(O(S, I, R, D) - O(S_1, I_1, R_1, D_1), (S - S_1, I - I_1, R - R_1, D - D_1)),$$
$$(5.95)$$

where the beta inner product is provided in the following definition.

Definition 5.5.1 (See [129]). Let h and J be two function defined on $[0, a]$. We suppose that $h \times J$ is beta integrable then, the beta inner product is defined as

$$A(h, J) = \int_0^a \left(t + \frac{1}{\Gamma(\beta)} \right)^{\beta-1} h(t)\overline{J(t)} \, dt. \qquad (5.96)$$

$\square$

Remark 2 (See [114]). We can notice that if a is a finite number, then the beta inner product can be bounded by the inner product as follows

$$A(h, J) = \int_0^a \left(t + \frac{1}{\Gamma(\beta)} \right)^{\beta-1} h(t)\overline{J(t)} \, dt \le \int_0^a h(t)\overline{J(t)} \, dt$$

$$b_\beta = \left(a + \frac{1}{\Gamma(\beta)} \right)^{\beta-1}. \qquad (5.97)$$

Therefore, using the above remark, we obtain the following:

$$G = A(O(S, I, R, D) - O(S_1, I_1, R_1, D_1), (S - S_1, I - I_1, R - R_1, D - D_1))$$
$$\le b_\beta(O(S, I, R, D) - O(S_1, I_1, R_1, D_1), (S - S_1, I - I_1, R - R_1, D - D_1)).$$
$$(5.98)$$

Our next concern now is to evaluate the inner product

$$(O(S, I, R, D) - O(S_1, I_1, R_1, D_1), (S - S_1, I - I_1, R - R_1, D - D_1))$$

$$= \begin{cases} cS(t)I(t) - cS_1(t)I_1(t) + f(R(t) - R_1(t)) - l(S(t) - S_1(t) \\ cS(t)I(t) - cS_1(t)I_1(t) - (f + h)(I(t) - I_1(t)) - h(S(t) - S_1(t) \\ h(I(t) - I_1(t)) - f(R(t) - R_1(t)) \\ f(I(t) - I_1(t)) + l(S(t) - S_1(t)). \end{cases}$$

$$(5.99)$$

We shall now evaluate

$$cS(t)I(t) - cS_1(t)I_1(t) + f(R(t) - R_1(t)) - l(S(t) - S_1(t))$$
$$\leq c\|S - S_1\|\|S_1I_1 - SI\| + f\|R - R_1\|\|S - S_1\| + |h - l|\|S - S_1\|^2.$$
$$(5.100)$$

Since the spread is taking place in a finite space and time, the solutions and the parameters involved in equation equation (5.70) are bounded. Therefore, we can find some positive parameters K_1, K_2, K_3, such that

$$\|R - R_1\| \leq K_1, \quad \|S_1 - S\| \leq K_2, \quad \text{and} \quad \|S_1I_1 - SI\| \leq K_3. \quad (5.101)$$

Therefore, replacing this in Equation (5.101), we obtain the following inequality

$$(cS(t)I(t) - cS_1(t)I_1(t) + f(R(t) - R_1(t)) - l(S(t) - S_1(t)), S(t) - S_1(t))$$
$$\leq (cK_1 + fK_3 + |h - l|)\|S - S_1\| = M_1\|S - S_1\|. \quad (5.102)$$

Also

$$(cS(t)I(t) - cS_1(t)I_1(t) - (f+h)(I(t) - I_1(t)) - h(S(t) - S_1(t), (S(t) - S_1(t))) \leq M_2\|I - I_1\|, \quad (5.103)$$

$$(h(I(t) - I_1(t)) - f(R(t) - R_1(t)), R(t) - R_1(t)) \leq M_3\|R - R_1\|$$
$$(f(I(t) - I_1(t)) + l(S(t) - S_1(t)), D(t) - D_1(t)) \leq M_4\|D - D_1\|. \quad (5.104)$$

Now, replacing Equations (5.104), (5.103), and (5.102) into Equation (5.105), we obtain

$$(O(S, I, R, D) - O(S_1, I_1, R_1, D_1), (S - S_1, I - I_1, R - R_1, D - D_1)) = \begin{cases} M_1\|S - S_1\| \\ M_2\|I - I_1\| \\ M_3\|R - R_1\| \\ M_4\|D - D_1\|. \end{cases}$$
$$(5.105)$$

The above leads to

$$G \leq b_\beta \begin{cases} M_1\|S - S_1\| \\ M_2\|I - I_1\| \\ M_3\|R - R_1\| \\ M_4\|D - D_1\|. \end{cases} \quad (5.106)$$

Following the same line of ideas, we can find a positive vector $N(N_1, N_2, N_3, N_4)$ such that for all vector $W(W_1, W_2, W_3, W_4) \in Z$

$$G \le b_\beta \begin{cases} N_1 \|S - S_1\| \|W_1\| \\ N_2 \|I - I_1\| \|W_2\| \\ N_3 \|R - R_1\| \|W_3\| \\ N_4 \|D - D_1\| \|N_4\|. \end{cases} \tag{5.107}$$

Now, putting together Equations (5.107) and (5.106), we complete the proof of Theorem 5.5.2.

Theorem 5.5.3 (See [114]). *Taking into account the initial conditions of Equation (5.91), and assuming that Equation (5.91) has an exact solution $U \ne$ for which the special solution converges to for large number n, then there is only one unique special solution for Equation (5.91) while using the new variational iteration method.*

Proof. Assume that I is the exact solution of system (5.91), let T and T_1 be two difference special solutions of system and converge to $U \ne 0$ for some large number n and m while using the decomposition method, then using Theorem 5.5.2, we have the following inequality

$$(O(t_1, t_2, t_3, t_4) - O(t_{11}, t_{12}, t_{13}, t_{14})) \le b_\beta N \|T - T_1\| = b_\beta \|T - U + U - T_1\|$$
$$\le b_\beta N(\|U - T_1\| + \|T - U\|) \|U\|. \tag{5.108}$$

Nevertheless, subsequently T and T_1 convergence to U for large number n and m. We can then find a small positive parameter ϵ, such that

$$\|U - T_1\| < \frac{\epsilon}{2b_\beta N \|U\|}, \quad \text{for } n,$$

$$\|U - T\| < \frac{\epsilon}{2b_\beta N \|U\|}, \quad \text{for } m. \tag{5.109}$$

Nonetheless, using the topology knowledge, we have

$$N \|T - T_1\| \|U\| = 0. \tag{5.110}$$

Since $U \ne 0$ and $N \ne 0$, then $\|T - T_1\| = 0$ implying $T = T_1$. This shows the uniqueness of the special solution. □

Figures 5.8–5.10 show the numerical simulations of future prediction for S, L, I, R, and V populations for different values of beta (Figure 5.11).

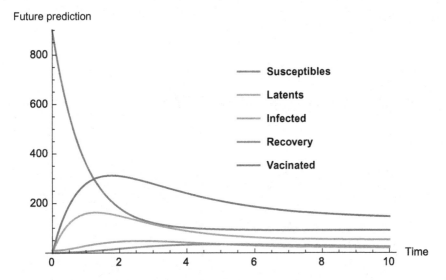

Figure 5.8 Prediction of the model for beta = 0.85.

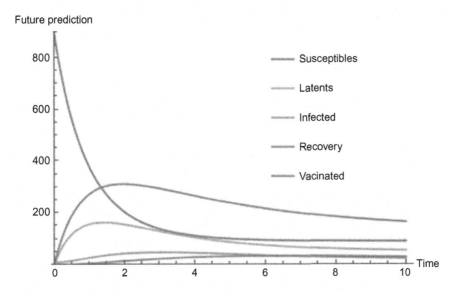

Figure 5.9 Prediction of the model for $\beta = 0.9$.

Future prediction

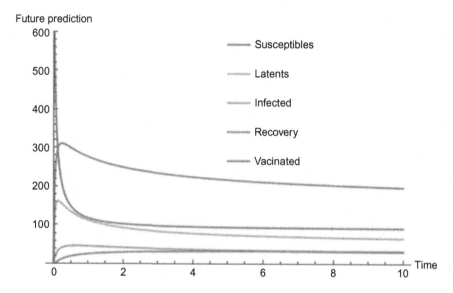

Figure 5.10 Prediction of the model for $\beta = 0.04$.

Future prediction

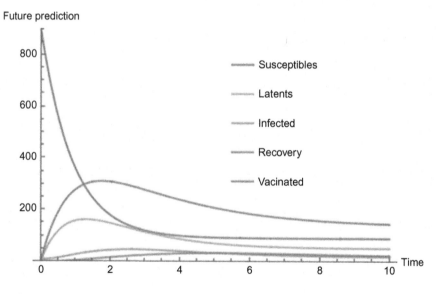

Figure 5.11 Prediction of the model for $\beta = 1$.

5.6 MODEL OF EBOLA HEMORRHAGIC FEVER

Ebolavirus disease (EVD; also Ebola hemorrhagic fever [EHF]), or simply Ebola, is a disease of humans and other primates caused by Ebolaviruses. Signs and symptoms typically start between 2 days and 3 weeks after contracting the virus with a fever, sore throat, muscle pain, and headaches. Vomiting, diarrhea, and rash usually follow, along with decreased function of the liver and kidneys. At this time some people begin to bleed both internally and externally [129, 130]. The disease has a high risk of death, killing between 25% and 90% of those infected, with an average of about 50% [130]. This is often due to low blood pressure from fluid loss, and typically follows 6–16 days after symptoms appear [131]. The virus spreads by direct contact with body fluids, such as blood, of an infected human or other animals [129, 130]. This may also occur through contact with an item recently contaminated with bodily fluids spread of the disease through the air between primates, including humans, has not been documented in either laboratory or natural conditions [130, 132]. Semen or breast milk of a person after recovery from EVD may still carry the virus for several weeks to months [130, 132]. Fruit bats are believed to be the normal carrier in nature, able to spread the virus without being affected by it. Other diseases such as malaria, cholera, typhoid fever, meningitis, and other viral hemorrhagic fevers may resemble EVD. Blood samples are tested for viral RNA, viral antibodies, or for the virus itself to confirm the diagnosis. EVD in humans is caused by four of five viruses of the genus Ebolavirus. These four are Bundibugyo virus (BDBV), Sudan virus (SUDV), Ta Forest virus (TAFV), and one simply called Ebolavirus (EBOV, formerly Zaire Ebolavirus) [133, 134]. EBOV, species Zaire Ebolavirus, is the most dangerous of the known EVD-causing viruses, and is responsible for the largest number of outbreaks [132, 135–137]. The fifth virus, Reston virus (RESTV), is not thought to cause disease in humans, but has caused disease in other primates [133, 134]. All five viruses are closely related to Marburgviruses.

5.6.1 Signs and symptoms

The length of time between exposure to the virus and the development of symptoms (incubation period) is between 2 and 21 days [132], usually between 4 and 10 days [135]. However, recent estimates based on mathematical models predict that around 5% of cases may take more than 21 days to develop [132, 136]. Symptoms usually begin with a sudden influenza-like stage characterized by fatigue, fever, weakness, decreased appetite, muscle pain, joint pain, headache, and sore throat [132, 135–137]. The fever is usually higher than 38.3 °C (101 °F) [133]. This is often followed by vomiting, diarrhea, and abdominal pain [133, 136, 137]. Next, shortness of breath and chest pain may occur, along with swelling, headaches, and confusion [132, 135–137]. In about half of the cases, the skin may develop a maculopapular rash, a flat red area covered with small bumps, 5–7 days after symptoms begin [132, 135–137]. In some cases, internal and external bleeding may occur [132, 135–137]. This typically begins 5–7 days after the first symptoms [134]. All infected people show some decreased blood clotting [132, 135–137]. Bleeding from mucous membranes or from sites of needle punctures has been reported in 40–50% of cases [132]. This may cause vomiting blood, coughing up of blood, or blood in stool [132, 135–137]. Bleeding into the skin may create petechiae, purpura, ecchymoses, or hematomas (especially around needle injection sites) [132, 135–137]. Bleeding into the whites of the eyes may also occur. Heavy bleeding is uncommon; if it occurs, it is usually located within the gastrointestinal tract [133]. Recovery may begin between 7 and 14 days after the first symptoms [132, 135–137]. Death, if it occurs, follows typically 6–16 days from the first symptoms and is often due to low blood pressure from fluid loss [132, 135–137]. In general, bleeding often indicates a worse outcome, and blood loss may result in death [132, 135–137]. People are often in a coma near the end of life [132, 135–137]. Those who survive often have ongoing muscle and joint pain, liver inflammation, decreased hearing, and may have constitutional symptoms such as fatigue, continued weakness, decreased appetite, and difficulty returning to preillness weight [132, 135–137]. Additionally they develop antibodies against Ebola that last at least 10 years, but it is unclear if they are immune to repeated infections [132, 135–137]. If someone recovers from Ebola, they can no longer transmit the disease. Figure 5.12 shows the symptoms of Ebola.

5.6.2 Transmission

Between people, Ebola disease spreads only by direct contact with the blood or body fluids of a person who has developed symptoms of the disease [138]. Body fluids that may contain Ebola viruses include saliva, mucus, vomit, feces, sweat, tears, breast milk, urine, and semen [132, 133, 135]. The World Health Organization (WHO) states that only people who are very sick are able to spread Ebola disease in saliva, and the virus has not been reported to be transmitted through sweat. Most people spread the virus

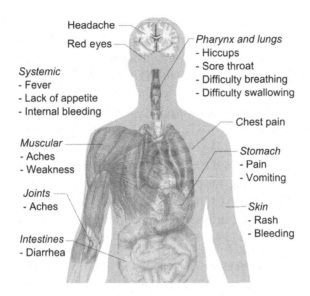

Headache
Red eyes

Pharynx and lungs
- Hiccups
- Sore throat
- Difficulty breathing
- Difficulty swallowing

Systemic
- Fever
- Lack of appetite
- Internal bleeding

Chest pain

Muscular
- Aches
- Weakness

Stomach
- Pain
- Vomiting

Joints
- Aches

Skin
- Rash
- Bleeding

Intestines
- Diarrhea

Figure 5.12 Ebola symptoms (see [134]).

through blood, feces, and vomit [128, 129, 131, 135, 138]. Entry points for the virus include the nose, mouth, eyes, open wounds, and cuts and abrasions [131]. Ebola may be spread through large droplets; however, this is believed to occur only when a person is very sick [138–141]. This can happen if a person is splashed with droplets [141–144]. Contact with surfaces or objects contaminated by the virus, particularly needles and syringes, may also transmit the infection [141–144]. The virus is able to survive on objects for a few hours in a dried state, and can survive for a few days within body fluids [141–144]. The Ebolavirus may be able to persist for up to 8 weeks in the semen after recovery, which could lead to infections via sexual intercourse [131, 132, 135, 136]. Ebola may also occur in the breast milk of women after recovery, and it is not known when it is safe to breastfeed again [141–144]. Otherwise, people who have recovered are not infectious [141–144]. The potential for widespread infections in countries with medical systems capable of observing correct medical isolation procedures is considered low [141–144]. Usually when someone has symptoms of the disease, they are unable to travel without assistance [141–144].

Dead bodies remain infectious; thus, people handling human remains in practices such as traditional burial rituals or more modern processes such as embalming are at risk [143, 144]. Sixty-nine percent of the cases of Ebola infections in Guinea during the 2014 outbreak are believed to have been contracted via unprotected (or unsuitably protected) contact with infected corpses during certain Guinean burial rituals [143, 144]. Healthcare workers treating people with Ebola are at the greatest risk of infection [141–144]. The risk increases when they do not have appropriate protective clothing such as masks, gowns, gloves, and eye protection, do not wear it properly, or handle contaminated clothing incorrectly [141–144]. This risk is particularly common in parts of Africa where the disease mostly occurs and health systems function poorly [141–144]. There has been transmission in hospitals in some African countries that reuse hypodermic needles [141–144]. Some healthcare centers caring for people with the disease do not have running water [141–144]. In the United States, the spread of the disease to two medical workers treating infected patients prompted criticism of inadequate training and procedures [141–144].

Human-to-human transmission of EBOV through the air has not been reported to occur during EVD outbreaks [132, 142], and airborne trans- mission has only been demonstrated in very strict laboratory conditions, and then only from pigs to primates, but not from primates to primates [141–144]. Spread of EBOV by water, or food other than bushmeat, has not been observed [141–144]. No spread by mosquitoes or other insects has been reported [141–144]. The apparent lack of airborne transmission among humans is believed to be due to low levels of the virus in the lungs and other parts of the respiratory system of primates, which is insufficient to cause new infections [141]. A number of studies examining airborne transmission broadly concluded that transmission from pigs to primates could happen without direct contact because, unlike humans and primates, pigs with EVD get very high Ebolavirus concentrations in their lungs, and not their bloodstream [141–144]. Therefore, pigs with EVD can spread the disease through droplets in the air or on the ground when they sneeze or cough [141–144]. By contrast, humans and other primates accumulate the virus throughout their body and specifically in their blood, but not very much in their lungs [141–144]. It is believed that this is the reason researchers have observed pig-to-primate transmission without physical contact, but no evidence has been found of primates being infected without actual contact, even in experiments where infected and uninfected primates shared the same air. Figure 5.13 shows the life cycle of the Ebolavirus. Bats are strongly implicated as both reservoirs and hosts for the Ebolavirus [141–144]. Of the five identified Ebolavirus subtypes, four are capable of human-to-human transmission. Initial infections in humans result from contact with an infected bat or other wild animal.

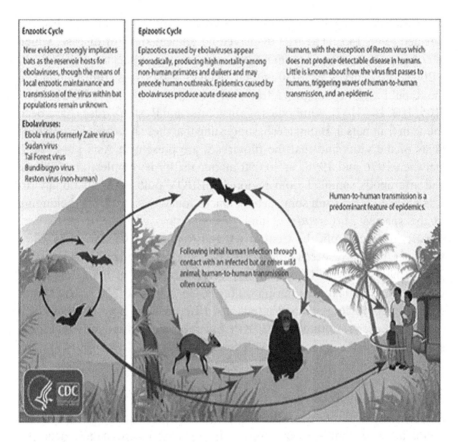

Enzootic Cycle

New evidence strongly implicates bats as the reservoir hosts for ebolaviruses, though the means of local enzootic maintainance and transmission of the virus within bat populations remain unknown.

Ebolaviruses:
Ebola virus (formerly Zaire virus)
Sudan virus
Taï Forest virus
Bundibugyo virus
Reston virus (non-human)

Epizootic Cycle

Epizootics caused by ebolaviruses appear sporadically, producing high mortality among non-human primates and duikers and may precede human outbreaks. Epidemics caused by ebolaviruses produce acute disease among humans, with the exception of Reston virus which does not produce detectable disease in humans. Little is known about how the virus first passes to humans, triggering waves of human-to-human transmission, and an epidemic.

Human-to-human transmission is a predominant feature of epidemics.

Following initial human infection through contact with an infected bat or other wild animal, human-to-human transmission often occurs.

Figure 5.13 Live cycle of Ebolavirus. From Centers for Disease Control and Prevention [134].

Strict isolation of infected patients is essential to reduce onward Ebolavirus transmission.

5.6.3 Host of Ebolavirus

The natural reservoir for Ebola has yet to be confirmed; however, bats are considered to be the most likely candidate species [143–145]. Three types of fruit bats (*Hypsignathus monstrosus*, *Epomops franqueti*, and *Myonycteris torquata*) were found to possibly carry the virus without getting sick [144, 145]. As of 2015, whether other animals are involved in its spread is not known [144, 145]. Plants, arthropods, and birds have also been considered possible viral reservoirs [129]. Bats were known to roost in the cotton factory in which the first cases of the 1976 and 1979 outbreaks were observed, and they have also been implicated in Marburg virus infections in 1975 and 1980 [144, 145]. Of 24 plant and 19 vertebrate species experimentally inoculated with EBOV, only bats became infected [144, 145]. The bats displayed no clinical signs of disease, which is considered evidence that these bats are a reservoir species of EBOV. In a 2002–2003 survey of 1030 animals including 679 bats from Gabon and the Republic of the Congo, 13 fruit bats were found to contain EBOV RNA [105, 109, 144–147]. Antibodies against Zaire and Reston viruses have been found in fruit bats in Bangladesh, suggesting that these bats are also potential hosts of the virus and that the filoviruses are present in Asia [144, 145]. Between 1976 and 1998, in 30,000 mammals, birds, reptiles, amphibians, and arthropods sampled from regions of EBOV outbreaks, no Ebolavirus was detected apart from some genetic traces found in six rodents (belonging to the species *Mus setulosus* and *Praomys*) and one shrew (*Sylvisorex ollula*) collected from the Central African Republic [105, 109, 144–147]. However, further research efforts have not confirmed rodents as a reservoir [105, 109, 144–147]. Traces of EBOV were detected in the carcasses of gorillas and chimpanzees during outbreaks in 2001 and 2003, which later became the source of human infections. However, the high rates of death in these species resulting from EBOV infection make it unlikely that these species represent a natural reservoir for the virus [105, 109, 146, 147].

5.6.4 Pathophysiology

Similar to other filoviruses, EBOV replicates very efficiently in many cells, producing large amounts of virus in monocytes, macrophages, dendritic cells, and other cells including liver cells, fibroblasts, and adrenal gland cells [144, 145]. Viral replication triggers the release of high levels of inflammatory chemical signals and leads to a septic state [144, 145]. EBOV is thought to infect humans through contact with mucous membranes or through skin breaks [105, 144, 147]. Once infected, endothelial cells (cells lining the inside of blood vessels), liver cells, and several types of immune cells such as macrophages, monocytes, and dendritic cells are the main targets of infection [105, 144, 147]. Following infection with the virus, the immune cells carry the virus to nearby lymph nodes where further reproduction of the virus takes place [147, 148]. From there, the virus can enter the bloodstream and lymphatic system and spread throughout the body [148]. Macrophages are the first cells infected with the virus, and this infection results in programmed cell death [144, 148, 149]. Other types of white blood cells, such as lymphocytes, also undergo programmed cell death, leading to an abnormally low concentration of lymphocytes in the blood [148, 149]. This contributes to the weakened immune response seen in those infected with EBOV [105, 144, 147]. Endothelial cells may be infected within 3 days after exposure to the virus [144, 145, 148, 149]. The breakdown of endothelial cells leading to blood vessel injury can be attributed to EBOV glycoproteins (GPs). This damage occurs due to the synthesis of Ebolavirus GP, which reduces the availability of specific integrins responsible for cell adhesion to the intercellular structure and causes liver damage, leading to improper clotting. The widespread bleeding that occurs in affected people causes swelling and shock due to loss of blood volume [148, 149]. The dysfunction in bleeding and clotting commonly seen in EVD has been attributed to increased activation of the extrinsic pathway of the coagulation cascade due to excessive tissue factor production by macrophages and monocytes [109, 144–146]. After infection, a secreted GP, small soluble glycoprotein (sGP or GP), is synthesized. EBOV replication overwhelms protein synthesis of infected cells and the host's immune defenses. The GP forms a trimeric complex, which tethers the virus to the endothelial cells. The sGP forms a dimeric protein that interferes with the signaling of neutrophils, another type of white blood cell, which enables the virus to evade the immune system by inhibiting early steps of neutrophil activation. The presence of viral particles and the cell damage resulting from viruses budding out of the cell causes the release of chemical signals (such as TNF-α, IL-6, and IL-8), which are molecular signals for fever and inflammation. Filoviral infection also interferes with proper functioning of the innate immune system [105, 148, 149]. EBOV proteins blunt the human immune system's response to viral infections by interfering with the cells' ability to produce and respond to interferon proteins such as interferon-alpha, interferon-beta, and interferon-gamma [105, 109, 144–146, 150]. The VP24 and VP35 structural proteins of EBOV play a key role in this interference. When a cell is infected with EBOV, receptors located in the cell's cytosol (such as RIG-I and MDA5), or outside of the cytosol (such as Toll-like receptor 3 (TLR3), TLR7, TLR8, and TLR9), recognize infectious molecules associated with the virus [148, 150]. On TLR activation, proteins including interferon regulatory factor 3 and interferon regulatory factor 7 trigger a signaling cascade that leads to the expression of type 1 interferons [148, 150]. The type 1 interferons are then released and bind to the IFNAR1 and IFNAR2 receptors expressed on the surface of a neighboring cell [148, 150]. Once interferon has bound to its receptors on the neighboring cell, the signaling proteins STAT1 and STAT2 are activated and move to the cell's nucleus [148, 150]. This triggers the

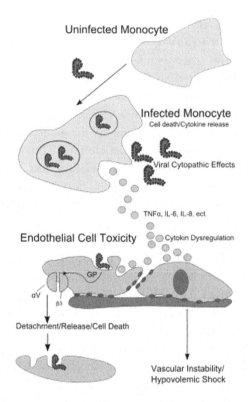

Uninfected Monocyte

Infected Monocyte
Cell death/Cytokine release

Viral Cytopathic Effects

TNFα, IL-6, IL-8, ect.

Endothelial Cell Toxicity   Cytokin Dysregulation

GP

αV   β3

Detachment/Release/Cell Death

Vascular Instability/
Hypovolemic Shock

Figure 5.14 Host immune responses to Ebolavirus and cell damage due to direct infection of monocytes and macrophages cause the release of cytokines associated with inflammation and fever. Infection of endothelial cells also induces a cytopathic effect and damage to the endothelial barrier that, together with cytokine effects, leads to the loss of vascular integrity. Transient expression of Ebolavirus GP in human umbilical vein endothelial cells or 293T cells causes a reduction of specific integrins (primary molecules responsible for cell adhesion to the extracellular matrix) and immune molecules on the cell surface. Cytokine dysregulation and virus infection may synergize at the endothelial surface, promoting hemorrhage and vasomotor collapse [149].

expression of interferon-stimulated genes, which code for proteins with antiviral properties (Figure 5.14).

5.6.5 Prevention of viral hemorrhagic fever

People who care for those infected with Ebola should wear protective clothing including masks, gloves, gowns, and goggles [150–154]. The US Centers for Disease Control (CDC) recommend that the protective gear leaves no skin exposed [150–154]. These measures are also recommended for those who may handle objects contaminated by an infected person's body fluids. In 2014, the CDC began recommending that medical personnel receive training on the proper suit-up and removal of personal protective equipment (PPE); in addition, a designated person, appropriately trained in biosafety, should be watching each step of these procedures to ensure they are done correctly [150–153]. In Sierra Leone, the typical training period for the use of such safety equipment lasts approximately 12 days [150–153]. The infected person should be in barrier isolation from other people [150–153]. All equipment, medical waste, patient waste, and surfaces that may have come into contact with body fluids need to be disinfected [150–153]. During the 2014 outbreak, kits were put together to help families treat Ebola in their homes, which included protective clothing as well as chlorine powder and other cleaning supplies [154, 155]. Education of those who provide care in these techniques, and the provision of such barrier-separation supplies, has been a priority for Doctors Without Borders [154, 155]. Ebolaviruses can be eliminated with heat (heating for 30–60 min at 60 °C or boiling for 5 min). To disinfect surfaces, some lipid solvents such as some alcohol-based products, detergents, sodium hypochlorite (bleach), or calcium hypochlorite (bleaching powder), and other suitable disinfectants may be used at appropriate concentrations [128, 132, 154, 155]. Education of the general public about the risk factors for Ebola infection and of the protective measures that individuals may take to prevent infection is recommended by the WHO [128, 129]. These measures include avoiding direct contact with infected people and regular hand washing using soap and water [150–153]. Bushmeat, an important source of protein in the diet of some Africans, should be handled and prepared with appropriate protective clothing and thoroughly cooked before consumption [128, 150–153]. Some research suggests that an outbreak of Ebola disease in wild animals killed for consumption may result in a corresponding human outbreak. Since 2003, such animal outbreaks have been monitored to predict and prevent Ebola outbreaks in humans [150–153].

If a person with Ebola dies, direct contact with the body should be avoided [153–155]. Certain burial rituals, which may have included making various direct contacts with a dead body, require reformulation in order to consistently maintain a proper protective barrier between the dead body and the living [150–155]. Social anthropologists may help find alternatives to traditional rules for burials [150–153]. Transportation crews are instructed to follow a certain isolation procedure should anyone exhibit symptoms resembling EVD [150–153]. As of August 2014, the WHO did not consider travel bans to be useful in decreasing spread of the disease [150–153]. In October 2014, the CDC defined four risk levels used to determine the level of 21-day monitoring for symptoms and restrictions on public activities [153–155]. In the United States, the CDC recommends that restrictions on public activity, including travel restrictions, are not required for the following defined risk levels [153–155]:

1. having been in a country with widespread Ebola transmission and having no known exposure (low risk), or having been in that country for more than 21 days ago (no risk);
2. encounter with a person showing symptoms; but not within three feet of the person with Ebola without wearing PPE, and no direct contact of body fluids;
3. having had brief skin contact with a person showing symptoms of Ebola when the person was believed to be not very contagious (low risk) in countries without widespread Ebola transmission, direct contact with a person showing symptoms of the disease while wearing PPE (low risk); and
4. contact with a person with Ebola before the person was showing symptoms (no risk).

The CDC recommends monitoring for the symptoms of Ebola disease for those both at "low risk" and at higher risk [153–155]. In laboratories where diagnostic testing is carried out, biosafety level 4 (BSL-4)-equivalent containment is required [153–155]. Laboratory researchers must be properly trained in BSL-4 practices and wear proper PPE [153–155] as shown in Figure 5.15.

5.6.6 Mathematical model of EHF via beta-derivative

A mathematical model underpinning the spread of Ebola virus with beta-derivative in a given country was recently proposed by Abdon Atangana and Franck Emile Goufo in 2014 [157], and in this section we present their model. Let us consider a given country with a total number of populations M at a given time. Let us assume that the rate of death due to natural causes and other diseases is factored to be α. s, i, r, and d are to be rate of infection by Ebola, rate of recovery, rate of susceptibility, and rate of death from Ebola, respectively. In their work, Atangana and Goufo considered $S(t)$, $I(t)$, $R(t)$, and $D(t)$ to be the representation of the susceptible, the infected, the recovery, and the total death populations, respectively. Let i is the rate of infectious person recovery population that turned out to be vulnerable again at the rate s and finally the number of population that dies of natural causes

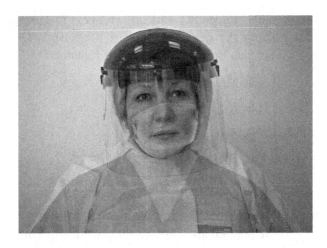

Figure 5.15 Senior Sister Donna Wood has been nursing for 29 years. She's one of the many medics from across Britain's National Health Service who are joining the UK's fight against Ebola in Sierra Leone (see [156]).

and other diseases at the rate α. The mathematical equation portraying this variation in respect to time is then provided by

$$_{0}^{A}D_{t}^{\beta}(S(t)) = -iS(t)I(t) + sR(t) - \alpha N. \tag{5.111}$$

To provide the model of an infectious population, we consider the fact that the total number of persons removed from the susceptible group can be mathematically expressed as $iS(t)I(t)$. Nonetheless, because of the introduction of medication, a number of individuals will be recover at the rate of r and also a number of infected persons will die at the rate d. The variation in time of the infectious population is then provided by the following mathematical model

$$_{0}^{A}D_{t}^{\beta}(I(t)) = iS(t)I(t) - dI(t) - rI(t). \tag{5.112}$$

Obviously the model describing the variation in time of recovery population is provided by

$$_{0}^{A}D_{t}^{\beta}(R(t)) = rI(t) - sR(t). \tag{5.113}$$

In the same way, we obtain the rate of change of death population as

$$_{0}^{A}D_{t}^{\beta}(I(t)) = dI(t) + \alpha N. \tag{5.114}$$

Now, putting together Equations (5.111), (5.112), (5.113) and Equation (5.114), we obtain the general model of Ebola spread, which was recently named "Atangana Beta Ebola System of Equation."

$$\begin{cases} {}_0^A D_t^\beta S(t) = -iS(t)I(t) + sR(t) - \alpha N \\ {}_0^A D_t^\beta I(t) = iS(t)I(t) - dI(t) - rI(t) \\ {}_0^A D_t^\beta R(t) = rI(t) - sR(t) \\ {}_0^A D_t^\beta D(t) = dI(t) + \alpha N. \end{cases} \tag{5.115}$$

5.6.7 Control of the disease via mathematical analysis

We present first some preliminaries for the study of stability of equilibrium points. For the global stability, we present first the Poincare–Bendixon theorem. For a two-dimensional system, bounded paths approach

1. an equilibrium point;
2. a limit cycle; or
3. a cycle graph.

Limit cycles must contain at least one equilibrium in their interior. Cyclic graphics are not possible from a stable equilibrium. The Bendixon–Dulac test applies

$$x'(t) = F(x, y)$$

$$y'(t) = G(x, t), \tag{5.116}$$

$(x, y) \in D$ simply connected, $F(x, y), G(x, y) \in C^1(D)$

$$\frac{\partial HF}{\partial x} + \frac{\partial HG}{\partial y}. \tag{5.117}$$

The sign is stable in D for some $H(x, y) \in C^1(D)$, implying that there is no periodic solution or cyclic graphs in D. Here $R_0 \leq 1$: $E_0 \in \partial$ is the only equilibrium point in T. There is no limit cycle in T. There is no cyclic graph in T. This implies that all paths in T approach E_0. $R_0 > 1$: E_0 is a saddle, $(S, 0)$ implies E_0 for $0 \leq S \leq 1$.

The General Method of Lyapunov:

$$V : U \subset \mathcal{R}^n \to \mathcal{R}; \quad \bar{0} \in U, \quad V \in C^1(U) \tag{5.118}$$

is positive definite on U:

1. $V(\overline{0}) = 0$
2. $V(\overline{x}) > 0, \overline{x} \neq \overline{0} \in U.$ V is negative definite if $-V$ is positive definite.

$$\overline{x}(t)' = f(\overline{x})$$

$$\overline{x} = (x_1, x_2, x_3, \ldots, x_n) \in \mathcal{R}^n$$

$$f(\overline{x}) = (f_1(\overline{x}), f_2(\overline{x}), f_3(\overline{x}), \ldots, f_n(\overline{x}) \in C^1(D)). \tag{5.119}$$

The orbital derivative of V along the trajectory $\overline{x}(t)$:

$$V(x(t)) = \sum_{i=1}^{n} \frac{\partial V(\overline{x}(t))}{\partial x_i} x_i'(t). \tag{5.120}$$

Theorem 5.6.1 (Lyapunov). *Let $\overline{0}$ an equilibrium point of $\overline{x}'$, V positive definite on a neighborhood U of $\overline{0}$.*

1. *If $V(\overline{x}) \leq 0$ for $\overline{x} \in U - \overline{0}$ implies $\overline{0}$ is stable.*
2. *If $V(\overline{x}) < 0$ for $\overline{x} \in U - \overline{0}$ implies $\overline{0}$ is asymptotically stable.*
3. *If $V(\overline{x}) > 0$ for $\overline{x} \in U - \overline{0}$ implies $\overline{0}$ is unstable.*

V is a Lyapunov function, if V is positive definite and $V(\overline{x}) \leq 0$ in $M \subset \mathcal{R}^n$ with M an invariant set under the flow of $\overline{x}' = f(\overline{x})$, if for any $\overline{x}^0 \in M$, then the solution trajectories through $\overline{x}^0$ belongs to M for all $t \in \mathcal{R}$.

Theorem 5.6.2 (La Salle–Lyapunov). *Let V a $C^1(\mathcal{R}^n)$ real valued function, $U = (\overline{x} \in \mathcal{R}^n | V(\overline{x}) < k, k \in \mathcal{R})$ and $V(\overline{x}) \leq 0$. M the largest invariant set in $S = \overline{x} \in U | V(\overline{x} = 0)$. Then every path that starts in U and remains bounded approaches to M.*

A generalization of the Kermack and McKendrick model is used to introduce more realistic situations

$$\overline{z}' = \text{diag}(\overline{x})(\overline{e} + A\overline{A})\overline{b}\overline{z}$$

$$\mathcal{R}_+^n = (\overline{z} \in \mathcal{R}^n, z_i \geq, i = 1, \ldots, n). \tag{5.121}$$

1. $\overline{e} \in \mathcal{R}^n$, a constant vector;
2. $A = (a_{ij})$, a real constant matrix;
3. $\overline{b}(\overline{z}) = \overline{c} + B\overline{z}, \overline{c} \in \mathcal{R}_+^n$, $B = (b_{ij})$ a constant non-negative matrix with $b_{ii} = 0$;

4.

$$\Omega = \left(\bar{z} \in \mathcal{R}^n_+ \Big/ \sum_{i=1}^{n} z_i \leq 1 \right) \quad \text{or} \quad \Omega = \left(\bar{z} \in \mathcal{R}^n_+ \Big/ \sum_{i=1}^{n} z_i \geq 1 \right) \tag{5.122}$$

are positively invariant under the flow induced.

The vector field

$$F(\bar{z}) = \mathrm{diag}(\bar{z})(\bar{e} + A\bar{A}) + \bar{b}\bar{z} \in C^1(\Omega). \tag{5.123}$$

Consider

$$D^i = (\bar{z} \in \Omega/z_i = 0). \tag{5.124}$$

1. If $b_i(\bar{z})|_{D_i} = 0$ implies $F(\bar{z})|_{D_i} = 0$ implies D_i is positively invariant.
2. If $b_i(\bar{z}) > 0$ implies $F(\bar{z}) \cdot \bar{n}_i < 0$, $\bar{n}_i$ is the exterior normal to Ω in D_i implies $F(\bar{z})$ points inside Ω.
3. The fixed point theorem assures the existence of at least one equilibrium solution within Ω.
4. If $\bar{c}$ is positive definite, then our system has positive equilibrium point $\bar{z}^*$. Define

$$\Omega_+ = (\bar{z} \in \Omega/z_i > 0, \quad i = 1, \ldots, n). \tag{5.125}$$

A positive equilibrium $\bar{z} \in \Omega$ is called an endemic equilibrium.
5. If an endemic equilibrium $\bar{z}^*$ is globally asymptotically stable with respect to Ω_+, this implies that $\bar{z}$ is unique.

5.6.8 Analysis and validation

Here, we first show that if

$$^A_0 D^\beta_t S(t) + {}^A_0 D^\beta_t R(t) + {}^A_0 D^\beta_t I(t) + {}^A_0 D^\beta_t D(t) = 0 \tag{5.126}$$

then

$$S(t) + R(t) + I(t) + D(t) = 0. \tag{5.127}$$

Indeed, with the addition of all parts of Equation (5.115), we have

$$^A_0 D^\beta_t S(t) + {}^A_0 D^\beta_t R(t) + {}^A_0 D^\beta_t I(t) + {}^A_0 D^\beta_t D(t) = 0. \tag{5.128}$$

Nevertheless, using the linearity of the beta-derivative, we obtain

$$^A_0 D^\beta_t (S(t) + R(t) + I(t) + D(t)) = 0. \tag{5.129}$$

Application of beta-integral on both sides of the above yields

$$(S(t) + R(t) + I(t) + D(t)) = \text{constance.} \tag{5.130}$$

Now, taking t to be zero, we obtain constance $= N$. We consider Equation (5.132) to find the endemic equilibrium point. Thanks to the beta-derivative that allows us to have that, a beta-derivative of a constant is zero because the equilibrium points are obtained here by assuming that the solution of Equation (5.132) does not depend on the time; then

$$\begin{cases} 0 = -iS^*I^* + sR^* - \alpha N \\ 0 = iS^*I^* - dI^* - rI^* \\ 0 = rI^* - sR^* \\ 0 = dI^* + \alpha N. \end{cases} \tag{5.131}$$

After some manipulations, we arrive at the following equilibrium points [114]:

$$\begin{cases} I^* = \frac{-\alpha N}{d} \\ R^* = \frac{\alpha N}{sd} \\ S^* = \frac{d+r}{i}. \end{cases} \tag{5.132}$$

According to a survey done in 2014, it was revealed that Ebola has a high rate of mortality with a maximum of 90% [149]. This implies that the recovery rate is very small, and then the existence conditions are true and are conformable to the real-world situation. We shall make use of the Jacobian method to find the Eigen-value associate of the endemic equilibrium points. The Jacobian matrices associate is given as:

$$J = \begin{bmatrix} 0 & -(d+r) & s \\ 0 & -i-d-r & 0 \\ 0 & r & -s \end{bmatrix}. \tag{5.133}$$

Nevertheless, at the free disease $\left(\frac{d+r}{i}, 0, 0\right)$, we have the following matrix:

$$J = \begin{bmatrix} -iI & -iS & s \\ -iI & -i-d-r & 0 \\ 0 & r & -s \end{bmatrix}. \tag{5.134}$$

To find the Eigen-value associate, we solve the following equation:

$$\det(J - \lambda I) = \begin{bmatrix} -\lambda & -(d+r) & s \\ 0 & -i-d-r-\lambda & 0 \\ 0 & r & -s-\lambda \end{bmatrix} = 0. \tag{5.135}$$

And the solutions of the above equation are given as:

$$\lambda = 0$$
$$\lambda = -2(r+d)$$
$$\lambda = -s. \tag{5.136}$$

To have a clear idea of the stability of the disease-free equilibrium point, one can use the Lyapunov approach described above. We shall present the derivation of the solution in the next section.

5.6.9 Derivation of the solution via the iterative method

Since the system is nonlinear, some analytical techniques such as Laplace transform, Fourier transform, and Green function will not be suitable for this case. Suitable methods for nonlinear equations have been documented in the literature, for instance, homotopy perturbation method and its derivatives [106, 158, 159] and variational iteration method and its derivatives [160–162]. However, in this work, we shall use the homotopy decomposition method. In this method, we first apply the inverse operator of ${}^A_0 D^\beta_t$ which is the beta-integral operator on both sides of system (5.132), to obtain,

$$\begin{cases} S(t) = S(0) + {}^A_0 I^\beta_t (-iS(t)I(t) + sR(t) - \alpha N), \\ I(t) = I(0) + {}^A_0 I^\beta_t (iS(t)I(t) - dI(t) - rI(t)), \\ R(t) = R(0) + {}^A_0 I^\beta_t (rI(t) - sR(t)), \\ D(t) = D(0) + {}^A_0 I^\beta_t (dI(t) + \alpha N). \end{cases} \tag{5.137}$$

The next move in this method is to assume that, since the system is nonlinear, the solutions can be obtained in series as:

$$\begin{cases} S(t) = \lim_{p \to 1} \sum_{n=0}^{\infty} p^n S_n(t), \\ I(t) = \lim_{p \to 1} \sum_{n=0}^{\infty} p^n I_n(t), \\ R(t) = \lim_{p \to 1} \sum_{n=0}^{\infty} p^n R_n(t), \\ D(t) = \lim_{p \to 1} \sum_{n=0}^{\infty} p^n D_n(t). \end{cases} \tag{5.138}$$

However, introducing the above-proposed solution into Equation (5.153), also making use of the idea of homotopy and after we compare terms of the same power of p, we obtain the following:

$$p^0 : \begin{cases} S_0(t) = S(0), \\ I_0(t) = I(0), \\ R_0(t) = R(0), \\ D_0(t) = D(0), \end{cases} \tag{5.139}$$

$$p^1 : \begin{cases} S_1(t) = {}_0^A I_t^\beta \left(-iS_0(t)I_0(t) + sR_0(t) - \alpha N \right), \\ I_1(t) = {}_0^A I_t^\beta \left(iS_0(t)I_0(t) - dI_0(t) - rI_0(t) \right), \\ R_1(t) = {}_0^A I_t^\beta \left(rI_0(t) - sR_0(t) \right), \\ D_1(t) = {}_0^A I_t^\beta \left(dI_0(t) + \alpha N \right). \end{cases} \quad (5.140)$$

$$p^2 : \begin{cases} S_2(t) = {}_0^A I_t^\beta \left(-iS_0(t)I_1(t) - iS_1(t)I_0(t) + sR_1(t) \right), \\ I_2(t) = {}_0^A I_t^\beta \left(iS_0(t)I_1(t) + iS_1(t)I_0(t) - rI_1(t) \right), \\ R_2(t) = {}_0^A I_t^\beta \left(rI_1(t) - sR_1(t) \right), \\ D_2(t) = {}_0^A I_t^\beta \left(dI_1(t) + \alpha N \right). \end{cases} \quad (5.141)$$

$$p^n : \begin{cases} S_n(t) = {}_0^A I_t^\beta \left(-\sum_{j=0}^{n-1} S_j I_{n-j-1} + sR_{n-1}(t) \right), \\ I_n(t) = {}_0^A I_t^\beta \left(\sum_{j=0}^{n-1} S_j I_{n-j-1} - rI_{n-1}(t) \right), \\ R_n(t) = {}_0^A I_t^\beta \left(rI_{n-1}(t) - sR_{n-1}(t) \right), \\ D_n(t) = {}_0^A I_t^\beta \left(dI_{n-1}(t) + \alpha N \right). \end{cases} \quad (5.142)$$

One of the important parts of any iteration method is to prove the uniqueness and the convergence of the method; we are going to show the analysis underpinning the convergence and the uniqueness of the proposed method for the general solution for $p = 1$.

Theorem 5.6.3 (See [157]). *Assume that A and B are Banach spaces and $L : A \to B$ is contraction nonlinear mapping. If the progression engendered by the one-dimensional homotopy decomposition method is, without loss of generality, regarded as:*

$$I_n(t) = L(I_{n-1}) = \sum_{k=0}^{n-1} S_k(t), \quad n = 1, 2, 3, \ldots, \quad (5.143)$$

then the following statements hold [157]:

(a)

$$\|I_n(t) - I(t)\| \le \rho^n \|I(0) - I(t)\|, \text{ with } 0 < \rho < 1; \quad (5.144)$$

(b) *For any $n > 0$, $I_n(t)$ is always in the neighborhood of the exact solution;*
(c) $\lim_{n \to \infty} I_n(t) = I(t)$.

Proof. We show the proof of (a) by employing the induction technique on the natural number n. Therefore, when $n = 1$, we have the following:

$$\|I_1(t) - I(t)\| = \|L(I_0(t) - I(t))\|. \quad (5.145)$$

However, by hypothesis, we establish that L has a fixed point, which is the exact solution. Because if $I(t)$ is the exact solution, then

$$I(t) = I_\infty(t) = L\left(\sum_{n=0}^{\infty-1} I_n(t)\right) = L\left(\sum_{n=0}^{\infty} I_n(t)\right) = \sum_{n=0}^{\infty} I_n(t). \qquad (5.146)$$

The above is correct because $\infty - 1 = \infty$. Thus, we officially have

$$I(t) = L(I(t)). \qquad (5.147)$$

Since L is a contractive nonlinear mapping, we shall have the following inequality:

$$\|L(I_0(t)) - L(I(t))\| \le \rho\|I_0(t) - I(t)\|, \quad 0 < \rho < 1. \qquad (5.148)$$

The property is then verified for $n = 1$. Assume that the hypothesis is verified for $n - 1$; we shall prove that it is also verified for n. However, at the level n, we have the following:

$$\|I_n(t) - I(t)\| = \|L(I_{n-1}(t)) - I(t)\| < \rho\|I_{n-1}(t) - I(t)\|. \qquad (5.149)$$

Using the fact that L is a nonlinear contractive mapping, we have the following:

$$\|L(I_{n-1}(t)) - L(I(t))\| < \rho\|L(I_{n-1}(t)) - L(I(t))\|. \qquad (5.150)$$

Furthermore, using the induction hypothesis, we arrive at:

$$\rho\|I_{n-1}(t) - I(t)\| < \rho\rho^{n-1}\|I_{n-1}(t) - I(t)\|. \qquad (5.151)$$

This completes the proof.   □

5.6.10 Numerical solutions

In this section, we present the numerical simulations of the model for different values of β. We consider the total number of population leaving the given country to be $N = 5000$, we assume that only 4000 persons are susceptible to be infected, and we predict the results after 12 months. The following algorithm is proposed for the numerical simulation purpose.

Algorithm 1.

The model depends on the order of the derivative; according to the prediction, when the beta is 0.95, our model predicts that, if we have initially 5000 then 4000 only are susceptible to be infected. We assume that the rate of infection is a constant and is $i = 0.01$, meaning that there is a low chance for a susceptible person to be infected by Ebola. We assume

$$\text{Input:} \begin{cases} S_0(t) = S(0), \\ I_0(t) = I(0), & \text{as initial input} \\ R_0(t) = R(0), \\ D_0(t) = D(0), \end{cases} \qquad (5.152)$$

a i—number of terms in the approximation.

b

$$\text{Input:} \begin{cases} S_{Ap}(t), \\ I_{Ap}(t), & \text{as approximate solution} \\ R_{Ap}(t), \\ D_{Ap}(t). \end{cases} \qquad (5.153)$$

$$\text{Step 1: Put Input:} \begin{cases} S_0(t) = S(0), \\ I_0(t) = I(0), \\ R_0(t) = R(0), \\ D_0(t) = D(0), \end{cases} \text{and} \begin{cases} S_{Ap}(t) \\ I_{Ap}(t), \\ R_{Ap}(t) \\ D_{Ap}(t) \end{cases} = \begin{cases} S_{Ap}(t) \\ I_{Ap}(t), \\ R_{Ap}(t) \\ D_{Ap}(t) \end{cases}$$

Step 2: for $i = 1$ to $n - 1$ do step 3, step 4, and step 5.

that an infected person will recover at the rate of $r = 0.4$; only 0.9% of the total population are likely to be susceptible. We assume that only 0.6% of infected persons will die. We do the simulation over a period of 12 months. Now for the $\beta = 0.95$ we have the following simulations: after a period of 12 months, approximately 2500 will die from approximately 3700 persons initially affected. Then, only approximately 1200 will recover (Figures 5.16–5.19). When the beta is 0.9, our model predicts that, if we have initially 5000 among whom 4000 only are susceptible to be infected. We assume that the rate of infection is a constant and is $i = 0.01$, meaning that there is a low chance for a susceptible person to be infected by Ebola. We assume that an infected person will recover at the rate of $r = 0.4$; only 0.9% of the total population are likely to be susceptible. We assume that only 0.6% of infected persons will die. We do the simulation over a period of 12 months. Now for the $\beta = 0.9$ we have the following simulations: after a period of 12 months, approximately 2550 will die from approximately 3750 persons initially affected. Then, only approximately 1200 will recover. When the beta is 0.5, our model predicts that, if we have initially 5000 among which 4000 only are susceptible to be infected. We assume that the

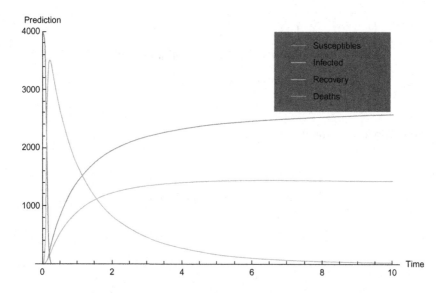

Figure 5.16 Numerical simulation of change in time of different populations for $\beta = 0.95$. From Simon Davis (DFID) [156].

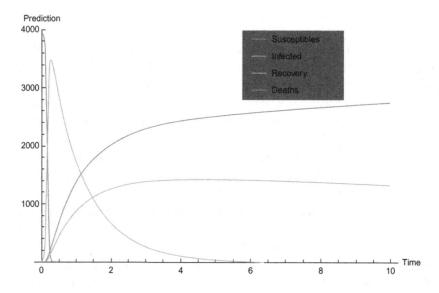

Figure 5.17 Numerical simulation of change in time of different populations for $\beta = 0.9$.

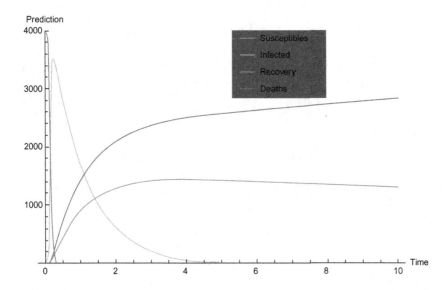

Figure 5.18 Numerical simulation of change in time of different populations for $\beta = 0.5$.

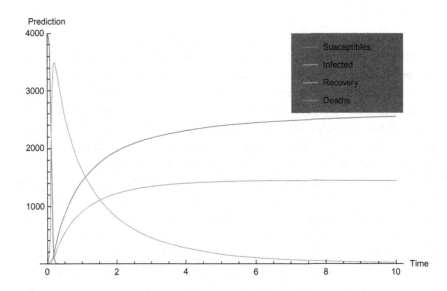

Figure 5.19 Numerical simulation of change in time of different populations for $\beta = 0.4$.

$$
\begin{cases}
S_n(t) = {}_0^A I_t^\beta \left(-\sum_{j=0}^{n-1} S_j I_{n-j-1} + sR_{n-1}(t) \right), \\
I_n(t) = {}_0^A I_t^\beta \left(\sum_{j=0}^{n-1} S_j I_{n-j-1} - rI_{n-1}(t) \right), \\
R_n(t) = {}_0^A I_t^\beta \left(rI_{n-1}(t) - sR_{n-1}(t) \right), \\
D_n(t) = {}_0^A I_t^\beta \left(dI_{n-1}(t) + \alpha N \right).
\end{cases}
\tag{5.154}
$$

Step 3:

$$
\begin{cases}
a_{n+1}(t) = a_n(t) + S_{Ap}(t), \\
b_{n+1}(t) = b_n(t) + I_{Ap}(t), \\
c_{n+1}(t) = c_n(t) + R_{Ap}(t), \\
d_{n+1}(t) = d_n(t) + D_{Ap}(t).
\end{cases}
\tag{5.155}
$$

Step 4:

$$
\begin{cases}
S_{Ap}(t) = a_{n+1}(t) + S_{Ap}(t), \\
I_{Ap}(t) = b_{n+1}(t) + I_{Ap}(t), \\
R_{Ap}(t) = c_{n+1}(t) + R_{Ap}(t), \\
D_{Ap}(t) = d_{n+1}(t) + D_{Ap}(t).
\end{cases}
\tag{5.156}
$$

Stop.

rate of infection is a constant and is $i = 0.01$, meaning that there is a low chance for a susceptible person to be infected by Ebola. We assume that an infected person will recover at the rate of $r = 0.4$; only 0.9% of the total population are likely to be susceptible. We assume that only 0.6% of infected persons will die. We do the simulation over a period of 12 months. Now for the $\beta = 0.5$ we have the following simulations: after a period of 12 months, approximately 2700 will die from approximately 3780 persons initially affected. Then, only approximately 1300 will recover. When the beta is 0.4, our model predicts that, if we have initially 5000 among whom 4000 only are susceptible to be infected. We assume that the rate of infection is a constant and is $i = 0.01$, meaning that there is a low chance for a susceptible person to be infected by Ebola. We assume that an infected person will recover at the rate of $r = 0.4$; only 0.9% of the total population are likely to be susceptible. We assume that only 0.6% of infected persons will die. We do the simulation over a period of 12 months. Now for the $\beta = 0.4$ we have the following simulations: after a period of 12 months, approximately 2800 will die from approximately 3800 persons initially affected. Then, only approximately 1200 will recover. The derivative used to model a real-world situation is very important. The classical derivative describes the change of rate, but it is an approximation of the real velocity of the object under study. The beta-derivative is the modification of the classical derivative that takes into account the time scale and also has a new parameter that can be considered as the fractional order. We have used the beta-derivative to model the spread of the fatal disease called Ebola that has killed many people in the West African countries including Nigeria, Sierra Leone, Guinea, and Liberia since December 2013. We did the investigation of the stable endemic points and presented the Eigen-values using the Jacobian method [157]. The homotopy decomposition method was used to solve the resulting system of equations [157]. The convergence of the method was presented and some numerical simulations were done for different values of beta. The simulations showed that our model is more realistic for all betas less than 9.5. The model revealed that, for a given population in a West African country, if there were no precautions of recovering, even if the total number of infected populations is very small, the entire population of that country would all die in a very short period of time [157]. Based on the prediction of this paper, we are calling for more research around this disease; in particular, we are calling for researchers to pay attention to finding a very good cure or better prevention, to reduce the risk of contamination [157].

BIBLIOGRAPHY

[1] T.A.A. Broadbent, M. Kline, The history of ancient Indian mathematics by C.N. Srinivasiengar, Math. Gaz. 52 (381) (1968) 307–308.

[2] R. Netz, K. Saito, N. Tchernetska, A new reading of method proposition 14: preliminary evidence from the Archimedes Palimpsest, I. SCIAMVS 2 (2001) 9–29.

[3] J.L. Berggren, Innovation and tradition in Sharaf Al-Din Al-Tusi's Muadalat, J. Am. Orient. Soc. 110 (2) (1990) 304–309.

[4] V.J. Katz, Ideas of calculus in Islam and India, Math. Mag. 68 (3) (1995) 163–174.

[5] J. Edwards, Differential Calculus, MacMillan and Co, London, 1892.

[6] I. Newton, The Principia: Mathematical Principles of Natural Philosophy, University of California Press, Berkeley, 1999, p. 974.

[7] E. Bodemann, Die Leibniz-Handschriften der Kniglichen ffentlichen Bibliothek zu Hannover, 1895, anastatic reprint, Georg Olms, Hildesheim, 1966.

[8] F. Cheesman, Isaac Newton's Teacher, first ed., vol. 115, Trafford Publishing, Victoria, BC, Canada, 2005, ISBN 1-4120-6700-6.

[9] G. Rodis-Lewis, Descartes' life and the development of his philosophy, in: J. Cottingham (Ed.), The Cambridge Companion to Descartes, vol. 22, Cambridge University Press, Cambridge, 1992, ISBN 978-0-521-36696-0.

[10] A.E. Bell, Christian Huygens and the Development of Science in the Seventeenth Century, Edward Arnold and Co, London, 1947.

[11] G. Mourlevat, Les Machines Arithmtiques De Blaise Pascal (In French), La Franaise d'Edition et d'Imprimerie, Clermont-Ferrand, 1988, p. 12.

[12] J.F. Scott, The Mathematical Work of John Wallis, D.D., F.R.S. (1616–1703), Chelsea Publishing Co., New York, NY, 1981, p. 18.

[13] M. Barany, Stuck in the middle: Cauchy's intermediate value theorem and the history of analytic rigor, Not. Am. Math. Soc. 60 (10) (2013) 1334–1338.

[14] S. Hawking, God Created the Integers, Running Press, Boston, 2005, pp. 814–815.

[15] J.J. O'Connor, E.F. Robertson, "Karl Theodor Wilhelm Weierstrass", School of Mathematics and Statistics, University of St Andrews, Scotland, 1998.

[16] B. Ross, A brief history and exposition of the fundamental theory of fractional calculus: fractional calculus and its applications, Lect. Notes Math. 457 (1975) 1–36.

[17] L. Debnath, A brief historical introduction to fractional calculus, Int. J. Math. Educ. Sci. Technol. 35 (4) (2004) 487–501.

[18] J.A.T. Machado, A.M.S.F. Galhano, J.J. Trujillo, On development of fractional calculus during the last fifty years, Scientometrics 98 (1) (2014) 577–582.

[19] A. Atangana, A. Secer, A note on fractional order derivatives and table of fractional derivatives of some special functions, Abstr. Appl. Anal. (2013) 1–8, Article ID 279681.

[20] M. Caputo, Linear models of dissipation whose Q is almost frequency independent. Part II, Geophys. J. Int. 13 (5) (1967) 529–539.

[21] A.A. Kilbas, H.M. Srivastava, J.J. Trujillo, Theory and applications of fractional differential equations, Elsevier, Amsterdam, The Netherlands, 2006.

[22] A. Erdlyi, On some functional transformations, Rendiconti del Seminario Matematico dell'Universit e del Politecnico di Torino 10 (1950–1951) 217–234.

[23] J. Hadamard, Essai sur l'tude des fonctions donnes par leur dveloppement de taylor, J. Pure Appl. Math. 4 (8) (1892) 101–186.

[24] M. Riesz, Intgrale De Riemann–Liouville Et Solution Invariantive Du Problme De Cauchy Pour L'quation Des Ondes, Comptes Rendus Du Congrs Internat. Des Math. Oslo, 1936, II, Oslo 1937.

[25] M. Davison, C. Essex, Fractional differential equations and initial value problems, Math. Sci. 23 (2) (1998) 108–116.

[26] C.F.M. Coimbra, Mechanics with variable-order differential operators, Ann. Phys. 12 (11–12) (2003) 692–703.

[27] G. Jumarie, Modified Riemann–Liouville derivative and fractional Taylor series of non-differentiable functions further results, Comput. Math. Appl. 51 (9–10) (2006) 367–376.

[28] F.H. Jackson, On q-functions and a certain difference operator, Trans. R. Soc. Edinb. 46 (1908) 253–281.

[29] R. Kanno, Representation of random walk in fractal space-time, Phys. A 248 (1998) 165–175.

[30] S. Umarov, S. Steinberg, Variable order differential equations with piecewise constant order-function and diffusion with changing modes, Z. Anal. Anwend. 28 (4) (2009) 431–450.

[31] A. Atangana, On the stability and convergence of the time-fractional variable order telegraph equation, J. Comput. Phys. (2015), URL http://dx.doi.org/10.1016/j.jcp.2014.12.043.

[32] B. Ross, S. Samko, Fractional integration operator of variable order in the holder spaces $H\Lambda(X)$, Int. J. Math. Math. Sci. 18 (4) (1995) 777–788.

[33] H.T.C. Pedro, M.H. Kobayashi, J.M.C. Pereira, C.F.M. Coimbra, Variable order modeling of diffusive-convective effects on the oscillatory flow past a sphere, J. Vib. Control 14 (9–10) (2008) 1659–1672.

[34] Y.L. Kobelev, L.Y. Kobelev, Y.L. Klimontovich, Statistical physics of dynamic systems with variable memory, Dokl. Phys. 48 (6) (2003) 285–289.

[35] A. Atangana, S.C.O. Noutchie, Stability and convergence of a time-fractional variable order Hantush equation for a deformable aquifer, Abstr. Appl. Anal. (2013) 1–8, Article ID 691060.

[36] A. Atangana, J.F. Botha, Generalized groundwater flow equation using the concept of variable order derivative, Boundary Value Problems 53 (2013) 1–11.

[37] A. Atangana, A. Kilicman, A novel integral operator transform and its application to some FODE and FPDE with some kind of singularities, Math. Probl. Eng. (2013) 1–7, Article ID 531984.

[38] A. Atangana, A. Kilicman, Analytical solutions of the space-time fractional derivative of advection dispersion equation, Math. Probl. Eng. (2013) 1–10, Article ID 8531279.

[39] A. Atangana, N. Bildik, The use of fractional order derivative to predict the groundwater flow, Math. Probl. Eng. (2013) 1–9, Article ID 543026.

[40] I. Podlubny, Fractional Differential Equations, vol. 198, Academic Press, New York, NY, USA, 1999.

[41] K.B. Oldham, J. Spanier, The Fractional Calculus, Academic Press, New York, NY, USA, 1974.

[42] A. Anatoly, J. Juan, M.S. Hari, Theory and Application of Fractional Differential Equations, vol. 204 of North-Holland Mathematics Studies, Elsevier, Amsterdam, The Netherlands, 2006.

[43] Y. Luchko, R. Groneflo, The Initial Value Problem for Some Fractional Differential Equations with the Caputo Derivative, Preprint Series A08-98, Fachbereich Mathematik Und Informatik, Freic Universitt, Berlin, Germany, 1998.

[44] O.P. Agrawal, Application of fractional derivatives in thermal analysis of disk brakes, Nonlinear Dynam. 38 (2004) 191–206.

[45] M.M. Meerschaert, C. Tadjeran, Finite difference approximations for fractional advection-dispersion flow equations, J. Comput. Appl. Math. 172 (1) (2004) 65–77.

[46] D.A. Benson, S.W. Wheatcraft, M.M. Meerschaert, Application of a fractional advection-dispersion equation, Water Resour. Res. 36 (6) (2000) 1403–1412.

[47] V. Daftardar-Gejji, H. Jafari, Adomian decomposition: a tool for solving a system of fractional differential equations, J. Math. Anal. Appl. 301 (2) (2005) 508–518.

[48] A. Cloot, J.F. Botha, A generalised groundwater flow equation using the concept of non-integer order derivatives, Water SA 32 (1) (2006) 55–78.

[49] M.D. Ortigueira, J.A.T. Machado, What is a fractional derivative?, J. Comput. Phys. (2015), doi: 10.1016/j.jcp.2014.07.019.

[50] J. Kestin, L.N. Persen, The transfer of heat across a turbulent boundary layer at very high Prandtl numbers, Int. J. Heat Mass Transfer 5 (1962) 355–371.

[51] H. Martin, The generalized Lvque equation and its practical use for the prediction of heat and mass transfer rates from pressure drop, Chem. Eng. Sci. 57 (16) (2002) 3217–3223.

[52] H. Schlichting, Boundary-Layer Theory, seventh ed., McGraw-Hill, New York, USA, 1979.

[53] A.H. Nayfeh, Perturbation Methods Wiley Classics Library, Wiley-Interscience, New York, 2000.

[54] F. Verhulst, Methods and Applications of Singular Perturbations: Boundary Layers and Multiple Timescale Dynamics, Springer, Dordrecht, 2005.

[55] A. Atangana, E.F.D. Goufo, Extension of matched asymptotic method to fractional boundary layers problems, Math. Probl. Eng. (2014) 1–7, Article ID 107535.

[56] M.A. Hammad, R. Khalil, Conformable fractional heat differential equation, Int. J. Pure Appl. Math. 94 (2) (2014) 215–221.

[57] M.A. Hammad, R. Khalil, Abels formula and Wronskian for conformable fractional differential equations, Int. J. Differ. Equat. Appl. 13 (3) (2014) 177–183.

[58] A. Atangana, E.F.D. Goufo, On the mathematical analysis of Ebola hemorrhagic fever: deathly infection disease in West African Countries, BioMed Res. Int. (2014) 1–7, Article ID 261383.

[59] R. Khalil, M.A. Horani, A. Yousef, M. Sababheh, A new definition of fractional derivative, J. Comput. Appl. Math. 264 (2014) 65–70.

[60] G.L. Caraffini, M. Galvani, Symmetries and exact solutions via conservation laws for some partial differential equations of mathematical physics, Appl. Math. Comput. 219 (2012) 1474–1484.

[61] G. Arfken, Integral transforms, in: Mathematical Methods for Physicists, third ed., Academic Press, Orlando, FL, 1985, pp. 794–864, ch. 16.

[62] Y.A. Brychkov, A.P. Prudnikov, Integral Transforms of Generalized Functions, Gordon and Breach, New York, 1989.

[63] H. Neidhardt, V.A. Zagrebnov, Linear non-autonomous Cauchy problems and evolution semigroups, Adv. Differential Equations 14 (3-4), (2009) pp. 289–340.

[64] H.S. Carslaw, J.C. Jaeger, Operational Methods in Applied Mathematics, Dover, New York, 1963.

[65] A. Erdlyi, M.F. Oberhettinger, F.G. Tricomi, Tables of Integral Transforms Based, in Part, on Notes Left by Harry Bateman and Compiled by the Staff of the Bateman Manuscript Project, vol. 2, McGraw-Hill, New York, 1954.

[66] S.G. Krantz, Transform theory, in: Handbook of Complex Variables, Birkhuser, Boston, MA, 1999, pp. 195–217, ch. 15.

[67] O.I. Marichev, Handbook of Integral Transforms of Higher Transcendental Functions: Theory and Algorithmic Tables, Ellis Horwood, Chichester, England, 1982.

[68] S.C. Oukouomi Noutchie, E.F. Doungmo Goufo, Global solvability of a continuous model for non-local fragmentation dynamics in a moving medium, Math. Probl. Eng. 2013 (2013) pp. 1–8, Article ID 320750.

[69] S.C.O. Noutchie, E.F.D. Goufo, On the honesty in non-local and discrete fragmentation dynamics in size and random position, ISRN Math. Anal. (2013).

[70] G.K. Watugala, Sumudu transform: a new integral transform to solve differential equations and control engineering problems, Int. J. Math. Educ. Sci. Technol. 24 (1) (1993) 35–43.

[71] S.R. Deans, The Radon Transform and Some of Its Applications, John Wiley and Sons, New York, 1983.

[72] G.G. Bilodeau, The Weierstrass transform and Hermite polynomials, Duke Math. J. 29 (ll) (1962) 293–308.

[73] Z.H. Khan, W.A. Khan, N-transform-properties and applications, NUST J. Eng. Sci. 1 (2008) 127–133.

[74] W. Rudin, Real and Complex Analysis, third ed., McGraw-Hill, Singapore, 1987, ISBN 0-07-100276-6.

[75] G.L. Lamb Jr., Introductory Applications of Partial Differential Equations with Emphasis on Wave Propagation and Diffusion, John Wiley and Sons, New York, NY, USA, 1995.

[76] A. Babakhani, R.S. Dahiya, Systems of multi-dimensional Laplace transforms and a heat equation, in: Proceedings of the 16th Conference on Applied Mathematics, Vol. 7 of Electronic Journal of Differential Equations, 2001, pp. 25–36.

[77] A. Atangana, A note on the triple Laplace transform and its applications to some kind of third-order differential equation, Abstr. Appl. Anal. (2013) 1–10, Article ID 769102.

[78] A.D. Polyanin, A.V. Manzhirov, Handbook of Integral Equations, CRC Press, Boca Raton, FL, USA, 1998.

[79] R.N. Bracewell, The Fourier Transform and Its Applications, third ed., McGraw-Hill, Boston, MA, USA, 2000.

[80] E.U. Condon, Immersion of the Fourier transform in a continuous group of functional transformations, Proc. Natl Acad. Sci. USA 23 (1937) 158–164.

[81] K. Watugala, Sumudu transform: a new integral transform to solve differential equations and control engineering problems, Int. J. Math. Educ. Sci. Technol. 24 (1) (1993) 35–43.

[82] S. Weerakoon, Application of Sumudu transform to partial differential equations, Int. J. Math. Educ. Sci. Technol. 25 (2) (1994) 277–283.

[83] A. Atangana, A. Kilicma, The use of Sumudu transform for solving certain nonlinear fractional heat-like equations, Abstr. Appl. Anal. (2013) 1–12, Article ID 737481.

[84] A. Atangana, Drawdown in prolate spheroidal-spherical coordinates obtained via Greens function and perturbation methods, Commun. Nonlinear Sci. Numer. Simul. 19 (5) (2014) 1259–1269.

[85] A. Atangana, On the singular perturbations for fractional differential equation, Scientific World J. (2014) 1–9, Article ID 752371.

[86] E. Yusufoglu, Variational iteration method for construction of some compact and noncompact structures of Klein-Gordon equations, Int. J. Nonlinear Sci. Numer. Simul. 8 (2) (2007) 153–158.

[87] G.-C. Wu, D. Baleanu, Variational iteration method for the Burgers' flow with fractional derivatives new Lagrange multipliers, Appl. Math. Model. 37 (9) (2013) 6183–6190.

[88] H. Jafari, V. Daftardar-Gejji, Solving a system of nonlinear fractional differential equations using adomian decomposition, J. Comput. Appl. Math. 196 (2) (2006) 644–651.

[89] H. Vazquez-Leal, A. Sarmiento-Reyes, Y. Khan, U. Filobello-Nino, A. Diaz-Sanchez, Rational biparameter homotopy perturbation method and Laplace-Pad coupled version, J. Appl. Math. (2012) 1–21, Article ID 923975.

[90] Y.-G. Wang, W.-H. Lin, N. Liu, A homotopy perturbation-based method for large deflection of a cantilever beam under a terminal follower force, Int. J. Comput. Methods Eng. Sci. Mechan. 13 (2012) 197–201.

[91] Z. Marda, Y. Khan, Singular initial value problem for a system of integro-differential equations, Abstr. Appl. Anal. (2012) 1–18, Article ID 918281.

[92] R. Hirota, J. Satsuma, Soliton solutions of a coupled Korteweg–De Vries equation, Phys. Lett. A 85 (8–9) (1981) 407–408.

[93] A. Atangana, A. Secer, The time-fractional coupled-Korteweg–De-Vries equations, Abstr. Appl. Anal. (2013) 1–8, Article ID 947986.

[94] A. Atangana, J.F. Botha, Analytical solution of the groundwater flow equation obtained via homotopy decomposition method, J. Earth Sci. Clim. Change 3 (2) (2012) 1–15.

[95] A.M. Wazwaz, The variational iteration method for rational solutions for Kdv, K(2,2), Burgers, and Cubic Boussinesq equations, J. Comput. Appl. Math. 207 (1) (2007) 18–23.

[96] Mustafa Inc., Numerical simulation of Kdv and mKdv equations with initial conditions by the variational iteration method chaos, Soliton. Fract. 34 (4) (2007) 1075–1081.

[97] M.A. Abdou, A.A. Soliman, Variational iteration method for solving burgers and coupled burgers equations, J. Comput. Appl. Math. 181 (2) (2005) 245–251.

[98] Y. Qing, B.E. Rhoades, T-stability of Picard iteration in metric spaces, Fixed Point Theory Appl. (2008) 1–4, Article ID 418971.

[99] C.-E. Froberg, Introduction to Numerical Analysis, Addison-Wesley, Reading, MA, USA, 1969.

[100] G.K. Watugala, Sumudu transform—a new integral transform to solve differential equations and control engineering problems, Math. Eng. Ind. 6 (4) (1998) 319–329.

[101] C.-M. Chen, F. Liu, I. Turner, V. Anh, A Fourier method for the fractional diffusion equation describing sub-diffusion, J. Comput. Phys. 227 (2) (2007) 886–897.

[102] Y.Q. Chen, K.L. Moore, Discretization schemes for fractional-order differentiators and integrators, IEEE Trans. Circuits Syst. I 49 (3) (2002) 363–367.

[103] S.B. Yuste, L. Acedo, An explicit finite difference method and a new Von Neumann-type stability analysis for fractional diffusion equations, SIAM J. Numer. Anal. 42 (5) (2005) 1862–1874.

[104] S.C. Oukouomi Noutchie, E.F. Doungmo Goufo, On the honesty in nonlocal and discrete fragmentation dynamics in size and random position, ISRN Math. Anal. 2013 (2013) 1–7, Article ID 908753.

[105] J.M. Morvan, V. Deubel, P. Gounon, E. Nakoun, P. Barrire, S. Murri, O. Perpte, B. Selekon, D. Coudrier, A. Gautier-Hion, M. Colyn, V. Volehkov, Identification of Ebola virus sequences present as RNA or DNA in organs of terrestrial small mammals of the Central African Republic, Microbes Infect. 1 (14) (1997) 1193–1201.

[106] Y. Tan, S. Abbasbandy, Homotopy analysis method for quadratic Riccati differential equation, Commun. Nonlinear Sci. Numer. Simul. 13 (3) (2008) 539–546.

[107] O. Ogbu, E. Ajuluchukwu, C.J. Uneke, Lassa fever in West African Sub-Region: an overview, J. Vector Borne Dis. 44 (1) (2007) 1–11.

[108] R. Hilfer, Application of Fractional Calculus in Physics, World Scientific, Singapore, 1999.

[109] K.J. Olival, A. Islam, M. Yu, S.J. Anthony, J.H. Epstein, S.A. Khan, S.U. Khan, G. Crameri, L.F. Wang, W.I. Lipkin, S.P. Luby, P. Daszak, Ebola virus antibodies in fruit bats, Bangladesh, Emerging Infect. Dis. 19 (2) (2013) 270–273.

[110] G.M. Duffield, AQTESOLV for *Windows Version 4.5 User's Guide*, HydroSOLVE, Inc., Reston, VA, (2007).

[111] Developed and Sold by Hydro SOLVE, Inc. The World's LEADING Aquifer Test Analysis Software Since (1989).

[112] C.V. Theis, The relation between the lowering of the piezometric surface and the rate and duration of discharge of a well using ground-water storage, Trans. Am. Geophys. Union 16 (1935) 519–524.

[113] A. Atangana, C. Unlu, New groundwater flow equation with its exact solution, (Ref. No: 40.716.150506), submitted to *Scientia Iranica*, (2015).

[114] A. Atangana, N. Bildik, The use of fractional order derivative to predict the groundwater flow, Math. Probl. Eng. (2013) 1–9, Article ID 543026.

[115] M.S. Hantush, C.E. Jacob, Non-steady radial flow in an infinite leaky aquifer, Am. Geophys. Union Trans. 36 (1) (1955) 95–100.

[116] M.S. Hantush, Aquifer tests on partially penetrating wells, J. Hyd. Div., Proc. Am. Soc. Civil Eng. 87 (5) (1961) 171–194.

[117] J.D. Frame, J.M. Baldwin, D.J. Gocke, J.M. Troup, Lassa fever, a new virus disease of man from West Africa. I. Clinical description and pathological findings, Am. J. Trop. Med. Hyg. 19 (4) (1970) 670–676.

[118] D. Werner (Ed.), Biological Resources and Migration, Springer, Berlin, 2004, p. 363.

[119] J. McCormick, A prospective study of the epidemiology and ecology of Lassa fever, J. Infect. Dis. 155 (1987) 437.

[120] J.K. Richmond, D.J. Baglole, Lassa fever: epidemiology, clinical features, and social consequences, BMJ 327 (7426) (2003) 1271–1275.

[121] S. Gnther, B. Weisner, A. Roth, T. Grewing, M. Asper, C. Drosten, P. Emmerich, J. Petersen, M. Wilczek, H. Schmitz, Lassa fever encephalopathy: Lassa virus in cerebrospinal fluid but not in serum, J. Infect. Dis. 184 (3) (2001) 345–349.

[122] K.M. Hastie, T. Liu, S. Li, L.B. King, N. Ngo, M.A. Zandonatti, V.L. Woods, Jr., J.C. de la Torre, E.O. Saphire, Crystal structure of the Lassa Virus Nucleoprotein–RNA complex reveals a gating mechanism for RNA binding, Proc. Natl. Acad. Sci. USA 108, 19365 (2011).

[123] R. Donaldson, The Lassa Ward, St. Martin's Press, New York, 2009.

[124] A. Atangana, A novel model for the Lassa hemorrhagic fever: deathly disease for pregnant women, Neural Comput. Appl. (2015), doi:10.1007/s00521-015-1860-9.

[125] O. Diekmann, J.A.P. Heesterbeek, J.A.J. Metz, On the definition and the computation of the basic reproduction ratio R_0 in models for infectious diseases in heterogeneous population, J. Math. Biol. 28 (1990) 365–382.

[126] Z. Feng, J.X. Velasco-Hernandez, Competitive exclusion in a vector-host model for the dengue fever, J. Math. Biol. 35 (1997) 523–544.

[127] O. Diekmann, J.A.P. Heesterbeek, J.A.J. Metz, On the definition and the computation of the basic reproduction ratio R_0 in models for infectious diseases in heterogeneous population, J. Math. Biol. 28 (1990) 365–382.

[128] S. Busenberg, C. Castillo-Chavez, A general solution of the problem of mixing subpopulations, and its application to risk- and age-structure epidemic models for the spread of AIDS, IMA J. Math. Appl. Med. Biol. 8 (1991) 1–29.

[129] A. Atangana, S.C.O. Noutchie, Model of breakbone fever via beta-derivatives, Biomed. Res. Int. (2014) 1–11, Article ID 523159.

[130] H. Feldmann, T.W. Geisbert, Ebola haemorrhagic fever, Lancet 377 (9768) (2011) 849–862.

[131] S.P. Fisher-Hoch, G.S. Platt, G.H. Neild, T. Southee, A. Baskerville, R.T. Raymond, G. Lloyd, D.I. Simpson, Pathophysiology of shock and hemorrhage in a fulminating viral infection Ebola, J. Infect. Dis. 152 (5) (1985) 887–894.

[132] Recommendations for breastfeeding and infant feeding in the context of Ebola, cdc.gov., September 19, 2014, Retrieved October 26, 2014.

[133] T. Hoenen, A. Groseth, D. Falzarano, H. Feldmann, Ebola virus: unravelling pathogenesis to combat a deadly disease, Trends Mol. Med. 12 (5) (2006) 206–215.

[134] Ebola Hemorrhagic Fever from Centers for Disease Control and Prevention, Retrieved January 28, 2014.

[135] M. Goeijenbier, J.J. van Kampen, C.B. Reusken, M.P. Koopmans, E.C. van Gorp, Ebola virus disease: a review on epidemiology, symptoms, treatment and pathogenesis, Neth. J. Med. 72 (9) (2014) 442–448.

[136] C.N. Haas, On the quarantine period for Ebola virus, PLOS Curr. Outbreaks (2014), doi:10.1371/currents.outbreaks.

[137] A. Khl, S. Phlmann, How Ebola virus counters the interferon system, Zoonoses Public Health 59 (2012) 116–131.

[138] K.M. Hastie, S. Bale, C.R. Kimberlin, E.O. Saphire, Hiding the evidence: two strategies for innate immune evasion by hemorrhagic fever viruses, Curr. Opin. Virol. 2 (2) (2012) 151–156.

[139] S. Mahanty, K. Hutchinson, S. Agarwal, M. McRae, P.E. Rollin, B. Pulendran, Cutting edge: impairment of dendritic cells and adaptive immunity by Ebola and Lassa viruses, J. Immunol. 170 (6) (2003) 2797–2801.

[140] D.G. McNeil. Jr., Ask Well: How Does Ebola Spread? How Long Can the Virus Survive?, The New York Times, Retrieved October 24, 2014.

[141] M. Chan, Ebola virus disease in West Africa—no early end to the outbreak, N. Engl. J. Med. 371 (13) (2014) 1183–1185.

[142] W.L. Irving, Ebola virus transmission, Int. J. Exp. Pathol. 76 (4) (1995) 225–226.

[143] J.P. Gonzalez, X. Pourrut, E. Leroy, Wildlife and emerging zoonotic diseases: the biology, circumstances and consequences of cross-species transmission, Curr. Topics Microbiol. Immunol. 315 (2007) 363–387, Ebolavirus and other filoviruses.

[144] R. Swanepoel, P.A. Leman, F.J. Burt, N.A. Zachariades, L.E. Braack, T.G. Ksiazek, P.E. Rollin, S.R. Zaki, C.J. Peters, Experimental inoculation of plants and animals with Ebola virus, Emerg. Infect. Dis. 2 (4) (1996) 321–325.

[145] H.M. Weingartl, C. Nfon, G. Kobinger, Review of Ebola virus infections in domestic animals, Dev. Biol. (Basel) 135 (2013) 211–218.

[146] K.B. Laupland, L. Valiquette, Ebola virus disease, Can. J. Infect. Dis. Med. Microbiol. 25 (3) (2014) 128–129.

[147] A. Groseth, H. Feldmann, J.E. Strong, The ecology of Ebola virus, Trends Microbiol. 15 (9) (2007) 408–416.

[148] S.R. Zaki, W. Shieh, P.W. Greer, C.S. Goldsmith, T. Ferebee, J. Katshitshi, F. Tshioko, M. Bwaka, R. Swanepoel, P. Calain, A.S. Khan, E. Lloyd, P. Rollin, T.G. Ksiazek, C.J. Peters, A novel immunohistochemical assay for the detection of Ebola virus in skin: implications for diagnosis, spread, and surveillance of Ebola hemorrhagic fever, J. Infect. Dis. 179 (1999) 36–47.

[149] N. Sullivan, Z.-Y. Yang, G.J. Nabel, Ebola virus pathogenesis: implications for vaccines and therapies, J. Virol. 77 (18) (2003) 9733–9737.

[150] A. Khl, S. Phlmann, How Ebola virus counters the interferon system, Zoonoses Public Health 59 (Suppl. 2) (2012) 116–131.

[151] World Health Organization and Centers for Disease Control and Prevention, Infection control for viral haemorrhagic fevers in the African Health Care Setting, December 1998, Retrieved October 20, 2014.

[152] World Health Organization, West Africa Ebola Virus Disease Update: Travel and Transport, International Travel and Health, 10 September (2014).

[153] Monitoring Symptoms and Controlling Movement to Stop Spread of Ebola, cdc.gov., October 27, 2014.

[154] Guidance on Personal Protective Equipment to Be Used by Healthcare Workers During Management of Patients with Ebola Virus Disease in U.S. Hospitals, Including Procedures for Putting on (Donning) and Removing (Doffing), cdc.gov., October 20, 2014.

[155] A. Nossiter, J. Kanter, Doctors Without Borders Evolves as It Forms the Vanguard in Ebola Fight, The New York Times, Retrieved October 12, 2014.

[156] Simon Davis (DFID) Donna Wood, Nurse and NHS Ebola, volunteer, November 17, 2014.

[157] A. Atangana, E.F.D. Goufo, On the mathematical analysis of Ebola hemorrhagic fever: deathly infection disease in West African Countries, Biomed. Res. Int. (2014) 1–7, Article ID 261383.

[158] J. He, Homotopy perturbation technique, Comput. Methods Appl. Mech. Eng. 178 (3–4) (1999) 257–262.

[159] M.A. Gondal, M. Khan, Homotopy perturbation method for nonlinear exponential boundary layer equation using Laplace transformation, He's polynomials and Pad technology, Int. J. Nonlinear Sci. Numer. Simul. 11 (12) (2010) 1145–1153.

[160] Z. Odibat, S. Momani, Numerical methods for nonlinear partial differential equations of fractional order, Appl. Math. Model. 32 (1) (2008) 28–39.

[161] G.C. Wu, D. Baleanu, Variational iteration method for the burgers' flow with fractional derivatives—new Lagrange multipliers, Appl. Math. Model. 37 (9) (2013) 6183–6190.

[162] I. Andrianov, J. Awrejcewicz, Construction of periodic solutions to partial differential equations with non-linear boundary conditions, Int. J. Nonlinear Sci. Numer. Simul. 1 (4) (2000) 327–332.